MIX
Papier aus verantwortungsvollen Quellen
Paper from responsible sources
FSC® C105338

Sven Bodo Wirsing

Separability within commutative and solvable associative algebras

Under consideration
of non-unitary algebras

With 401 exercises

Anchor Academic
Publishing

Wirsing, Sven Bodo: Separability within commutative and solvable associative algebras. Under consideration of non-unitary algebras. With 401 exercises, Hamburg, Anchor Academic Publishing 2018

Buch-ISBN: 978-3-96067-221-0
PDF-eBook-ISBN: 978-3-96067-721-5
Druck/Herstellung: Anchor Academic Publishing, Hamburg, 2018

Bibliografische Information der Deutschen Nationalbibliothek:
Die Deutsche Nationalbibliothek verzeichnet diese Publikation in der Deutschen Nationalbibliografie; detaillierte bibliografische Daten sind im Internet über http://dnb.d-nb.de abrufbar.

Bibliographical Information of the German National Library:
The German National Library lists this publication in the German National Bibliography. Detailed bibliographic data can be found at: http://dnb.d-nb.de

All rights reserved. This publication may not be reproduced, stored in a retrieval system or transmitted, in any form or by any means, electronic, mechanical, photocopying, recording or otherwise, without the prior permission of the publishers.

Das Werk einschließlich aller seiner Teile ist urheberrechtlich geschützt. Jede Verwertung außerhalb der Grenzen des Urheberrechtsgesetzes ist ohne Zustimmung des Verlages unzulässig und strafbar. Dies gilt insbesondere für Vervielfältigungen, Übersetzungen, Mikroverfilmungen und die Einspeicherung und Bearbeitung in elektronischen Systemen.

Die Wiedergabe von Gebrauchsnamen, Handelsnamen, Warenbezeichnungen usw. in diesem Werk berechtigt auch ohne besondere Kennzeichnung nicht zu der Annahme, dass solche Namen im Sinne der Warenzeichen- und Markenschutz-Gesetzgebung als frei zu betrachten wären und daher von jedermann benutzt werden dürften.

Die Informationen in diesem Werk wurden mit Sorgfalt erarbeitet. Dennoch können Fehler nicht vollständig ausgeschlossen werden und die Diplomica Verlag GmbH, die Autoren oder Übersetzer übernehmen keine juristische Verantwortung oder irgendeine Haftung für evtl. verbliebene fehlerhafte Angaben und deren Folgen.

Alle Rechte vorbehalten

© Anchor Academic Publishing, Imprint der Diplomica Verlag GmbH
Hermannstal 119k, 22119 Hamburg
http://www.diplomica-verlag.de, Hamburg 2018
Printed in Germany

Contents

Introduction   3

Notation   7

1 Separable algebras and the theorem of Wedderburn-Malcev   13
    1.1 Separable algebras 13
        1.1.1 Characterizations, properties and examples 13
        1.1.2 Group algebras and separability 15
        1.1.3 Matrix algebras of separable algebras 16
        1.1.4 Separable algebras, derivations and factor sets 17
            1.1.4.1 Derivations 17
            1.1.4.2 Factor sets 27
    1.2 Radical complements and the theorem of Wedderburn-Malcev   29
    1.3 Examples within the context of the theorem of Wedderburn-Malcev 30
        1.3.1 A counterexample and an example 30
        1.3.2 The algebras of upper and lower triangular matrices . 34
        1.3.3 Closed shapes and matrix algebras 37
    1.4 Invariant radical complements and Taft's theorem 38
    1.5 Connections to the theorems of Schur-Zassenhaus and Levi-Malcev 42
    1.6 Open-ended questions and exercises 45

2 Non-unitary algebras   57
    2.1 Adjunction of an unit 58
    2.2 The existence part 61
    2.3 The conjugacy part 66
    2.4 Cardinality of the set of radical complements 72
    2.5 Invariant radical complements and Taft's theorem 76
    2.6 Compatibility properties 77
    2.7 Top down calculation 80
    2.8 Open-ended questions and exercises 84

3	**Solvable algebras**	**93**
3.1	Basic properties	93
3.2	Connections to the associated Lie algebra	98
	3.2.1 A Lie characterization	98
	3.2.2 A symmetric bilinear form	103
	3.2.3 Group algebras	110
	3.2.4 Triangular matrices	111
3.3	Compatibilities	115
3.4	A summarizing example	119
3.5	Lower triangular matrices and solvable associative algebras	122
3.6	Open-ended questions and exercises	124
4	**Generalized quaternion algebras**	**137**
4.1	Definition and isomorphism	137
4.2	The case of a big nilradical	139
4.3	The case of characteristic not two	139
	4.3.1 The case within the literature	140
	4.3.2 One component is zero	140
4.4	The case of characteristic equal to 2	141
	4.4.1 One component is zero	143
	4.4.2 The third case	144
4.5	Exercises	148
5	**Commutative algebras**	**151**
5.1	Compatibility with the center	151
5.2	The subalgebra of fully separable elements	154
5.3	The context of the Wedderburn-Malcev theorem	157
5.4	A generalized Jordan decomposition	160
	5.4.1 The construction of the decomposition	160
	5.4.2 Properties of the generalized Jordan decomposition	167
5.5	The subalgebras of splitting and diagonalizable elements	172
5.6	Commutative group algebras	176
5.7	Non-unitary commutative associative algebras	180
5.8	Solvable algebras	188
5.9	Open-ended questions and exercises	192
5.10	Exercises	193
A	**About a theorem of Thorsten Bauer**	**215**
A.1	The proof	215
A.2	Exercises	217

B Proof of the Wedderburn-Malcev theorem for associative unitary algebras **219**
  B.1 The existence part 220
      B.1.1 The case of a zero nilradical - by using cohomology of algebras 220
      B.1.2 The induction argument 222
  B.2 The conjugacy part 222
      B.2.1 The case of a zero nilradical - by using cohomology of algebras 222
      B.2.2 The induction argument 224
  B.3 Open-ended questions and exercises 224

C Proof of Taft's theorem for associative unitary algebras **229**
  C.1 The existence part 229
      C.1.1 The case of a zero nilradical - cohomology of groups . 230
      C.1.2 The case of a zero nilradical - cohomology of algebras 233
      C.1.3 The case of a zero nilradical - derivations of algebra . 233
      C.1.4 The induction argument 234
  C.2 The conjugacy part 235
      C.2.1 The case of a zero nilradical - group cohomology . . . 235
      C.2.2 The case of a zero nilradical - cohomology of algebras 236
      C.2.3 The induction argument - cohomology of groups . . . 236
      C.2.4 The induction argument - direct computation 238
  C.3 Exercises 238

List of figures **241**

Bibliography **243**

Index **247**

For my

Super
Incredible
Splendid
Top
Exceptional
Reliable

Kerstin

and my

Bright
Remarkable
Outstanding
Terrific
Enormous
Helpful
Refreshing

Thorsten

Introduction

The truth is rarely pure and never simple. (Oscar Wilde)

Within the theory of associative algebras the nilradical and its factor algebra play an important role. The nilradical leads to the analysis of nilpotent and its factor algebra to the study of semisimple associative algebras. If the factor algebra by the nilradical of a finite-dimensional associative unitary algebra is separable, then the theorem of Wedderburn-Malcev ensures the existence of a subalgebra which is complementary to the nilradical. In other words, it is possible to lift the factor algebra by the nilradical into the algebra as a subalgebra. Furthermore, all such complements are conjugated under the action of the nilradical. An introduction to this topic is presented in chapter 1 of this work. In addition, we present the theorem of Taft for G-invariant radical complements (where G is a finite group acting on the algebra by auto- or anti-automorphism), include some examples of separable algebras, present the connection between separable algebras, derivations and factor sets, calculate the derivations of upper triangular matrices and present some examples and counterexamples within the context of the theorem of Wedderburn-Malcev.

Within the standard literature the theorem of Wedderburn-Malcev is proven for unital algebras. In some papers and books it is stated afterwards that every algebra can be embedded into an unital algebra, and thus the theorem is valid also for non-unital algebras (see e.g. [8], first paragraph of page 3). This idea will be analyzed in details in this work within chapter 2.

We study the so-called adjunction of an unit and the embedding of an algebra into this adjunction. By determining its nilradical and the factor algebra by its nilradical we are able to transfer the existence part of the theorem of Wedderburn-Malcev to non-unitary algebras. For non-unitary algebras the question arise in what way two complements are conjugated. For this, we introduce the well-known star group (also called quasi regular or circle group). By analyzing the star group and the connection to the adjunction of an unit we prove that all complements are conjugated under the action of the star group.

We analyze this action further by determining its stabilizer: it is the centralizer under the nilradical of a radical complement. This result is applied to the algebras of upper and lower triangular matrices over a field to determine the cardinality of the set of all radical complements.

The theorem of Taft for G-invariant radical complements is transferred to non-unitary algebras, too.

We proceed the chapter by analyzing compatibilities related to the theorem of Wedderburn-Malcev. The basic idea is to calculate radical complements of related structures (like subalgebras, ideals, factor algebras) based on radical complements of the entire algebra. For subalgebras, left and right ideals no meaningful compatibilities are provable in general (For the center one compatibility is proven within chapter 5.). For this, examples are presented. But we prove compatibilities for ideals and factor algebras by intersecting and factorizing radical complements in a natural way.

Chapter 2 is finalized by presenting algorithms for the determination of a radical complement. As a consequence we can calculate the decomposition for every element based on the nilradical and a radical complement. This decomposition can be used to calculate a decomposition based on every other radical complement by applying a transfer rule. The calculation of a decomposition for an element based on the decomposition of the entire algebra is called top-down calculation.

Two main topics are the guidelines of this work: the calculation of a radical complement and the presentation of an element as sum of a radical element and an element of a complement. In addition, the idea of compatibility is regarded as the third main topic within this work. For proving meaningful results related to this guidelines we will specialize the algebras to be analyzed.

Within chapter 3 we focus on so-called solvable associative algebras which are not or only little present in the basic literature. In the work [3] some deep insights in the theory of solvable associative algebras are proven. The analysis is motivated by Solomon's algebra and its connection to the representation theory of the symmetric groups. Solvable algebras are generalizing commutative algebras in a natural way. Thus, their analysis is also a basis for chapter 5 in this work.

Within the first section we prove that finite-dimensional associative solvable algebras are those algebras possessing a commutative factor algebra by its nilradical. Section two is dedicated to the result that solvable associative algebras (in uneven characteristic) are closely connected to the solvability of their associated Lie algebra. We use Cartan's criteria to characterize solvable associative algebras by a symmetric bilinear form. The connection to its associated Lie algebra motivated the question how the classes of solvabilities are connected. This topic is only analyzed for the algebras of upper triangular matrices over a field. We prove that both classes are identical

Introduction

and also equal to class of solvability of its group of units (for uneven characteristic).
By the idea of compatability we calculate radical complements for subalgebras of solvable algebras and prove that all semisimple subalgebras are separable (if the radical factor algebra of the underlying algebra is separable).
We finalize the chapter by presenting an example summarizing the results proven so far and by analyzing the importance of the lower and upper triangular matrices for all solvable associative algebras.

Chapter 4 is dedicated to algebras related to generalized quaternion algebras. We derive some examples for commutative algebras which are used within chapter 5. In addition, an algebra is presented possessing two non-conjugated radical complements. In the literature only very few examples related to this topic are existing. Finally, we classify the derived algebras up to isomorphism.

Our main questions are answered in details for commutative algebras within chapter 5 of this work. Commutative algebras possess exactly one radical complement. Standard examples of commutative algebras are centers of associative algebras. By using the idea of compatibility we describe the radical complement of the center based on the entire algebra: the intersection of every radical complement of the entire algebra with the center is exactly the radical complement of the center. As a consequence, we prove for solvable algebras that the intersection of all radical complements is exactly the radical complement of the center.
Afterwards, we turn our focus to the inner structure of commutative algebras. The unique radical complement is identified by the set of elements possessing a minimal polynomial which is squarefree and separable. These elements are called fully separable within this work. The decomposition topic is also solved by generalizing the Jordan decomposition. The latter decomposition is to be calculated by solving congruences within the algebra of polynomials. In the generalized version - applying not only to splitting endomorphism but also to separable ones - special divisions are to be done additionally. The set of diagonalizable and splitting elements are the connection to the well-known Jordan decompositions. Both sets are analyzed, and it is proven that they are subalgebras, too. The connection to the nilradical and the radical complement is presented for them.
The results are illustrated by using commutative group algebras and the algebras analyzed within chapter 4. In addition, we transfer all results to non-unitary associative algebras.
Finally, we answer our main topics for solvable associative algebras within chapter 5. The radical complements can be determined by using the set of fully separable elements. The generalized Jordan decomposition is used to

answer the decomposition question for the elements. We are able to generalize this calculation to solvable and to basic algebras. Thus, we are able to calculate a decomposition for an element without knowing a decomposition of the entire algebra. This approach is called bottom-up calculation. We can use the bottom-up calculation for describing a radical complement of the underlying algebra, too.

The appendix starts by analyzing a theorem within the work [3] of T. Bauer. The proof that solvable associative algebras can be characterized by algebras possessing solvable group of units is analyzed in details. The result is transferred to non-unitary algebras using again the star group and the adjunction of an unit. Thus, solvable associative algebras are characterized by possessing solvable groups of units and solvable associated Lie algebras. Afterwards we present proofs for the theorem of Wedderburn-Malcev and for Taft's theorem about G-invariant radical complements for unitary algebras using cohomology of algebras, groups and direct calculations.

Some applications are also transferred to the exercises at the end of each chapter. Some exercises are included enhancing the theory presented so far. In addition, at the beginning of each exercise series some open-ended topics are included which can be used by the reader – and also by the author – to do additional researches within this theory. The author has included some graphics – mostly so called Hasse diagrams – to visualize the main results of this work.

Notation

Numbers and sets

$\mathbb{P}$	the set of prime numbers
$\mathbb{N}$	the set of natural numbers without 0
$\mathbb{Z}$	the set of integers
$\mathbb{Q}$	the set of rational numbers
$\mathbb{R}$	the set of real numbers
$\mathbb{C}$	the set of complex numbers
$\mathbb{H}$	the set of real quaternions
$\underline{n}$	$\{a \mid a \in \mathbb{N}, a \leq n\}$
$[x]$	maximal integer less than or equal to x (Gauss bracket)
$A \times B$	the set of all pairs $(a; b)$, $a \in A$ and $b \in B$
M^n	the set of all n-tuples over M

Fields and polynomial rings

$(K; L)$	field extension with extension field L and basic field K
$K[t_1, ..., t_n]$	polynomial algebra in commutating variables $t_1, ..., t_n$ over K
$K(t_1, ..., t_n)$	field of fractions of $K[t_1, ..., t_n]$
$grad(f)$	degree of the polynomial $f \in K[t]$
$char(K)$	characteristic of the field K
(f)	notation for the K-ideal $fK[t]$ of $K[t]$
$K[a]$	smallest unitary K-subalgebra of a K-algebra containing a
$GF(p), p \in \mathbb{P}$	notation for the field $\mathbb{Z}/p\mathbb{Z}$
$halb(f)$	product of the pairwise irreducible divisors of the polynomial f
$max(f)$	greatest multiplicity of the irreducible divisors of the decomposition into irreducible polynomials of the polynomial f

Groups and magmas

G/U	the set of right cosets of a subgroup U of a group G
$G \times H$	the direct product of the groups G and H
$Stab_G(m)$	the stabilizer of an element m of a G-set
mG	the orbit of an element m of a G-set

$st(G)$	the solvable class of a solvable group G
$[g,h]$	the commutator of elements g,h of a group
G'	the commutator subgroup of a group G
$G^{(n)}$	the n-th derived subgroup of a group G
S_n	the symmetric group over $\underline{n}$
A_n	the alternating group over $\underline{n}$
D_{2n}	the dihedral group of order $2n$
Q_{4n}	the quaternion group of order $4n$
$O_p(G), p \in \mathbb{P}$	the intersection of all p-Sylow subgroups of a group G
$Aut(M)$	the set of all automorphism of a magma M

Spaces and matrices

$\langle T \rangle_K$	the K-linear span of a set T of vectors
Kv	the set $\langle v \rangle_K$
$End_K(V)$	the set of all K-endomorphism of a K-space V
$f(k)$	the inserting of k into the polynomial f
$dim_K(V)$	the K-dimension of a K-space V
$U \oplus_K W$	the inner direct sum of the K-subspaces U and W of a K-space
$V \otimes_K W$	the tensor product of the K-spaces V and W
$v \otimes w$	tensors of a tensor product
$ker(\alpha)$	kernel of the endomorphism α
$Im(\alpha)$	image of the endomorphism α
$\alpha \otimes \beta$	the tensor product of the K-linear functions α and β
tr	the trace function
$M_B(\alpha)$	the representing matrix of a K-linear function α based on a basis B
A_{ij}	the $(i;j)$-value of a matrix A
a_{ij}	the $(i;j)$-value of the matrix $A = (a_{ij})$
$K^{n \times m}$	$n \times m$-matrix space over K
$GL(n, K)$	the group of units of $K^{n \times n}$
$rad(f)$	the radical of a symmetric bilinear form f
$QA(K)$	the set of squares of a field K
$Pot(n, K)$	the set of n-th powers of a field K
τ	the transpose function on $K^{n \times n}$
$Aut(A)$	group of algebra automorphism of A
$Ant(A)$	set of anti-automorphism of A
$Der(A, M)$	set of derivations of a (A, A)-bimodule M
$Z^1(A, M)$	set of 1-cocycles of a (A, A)-bimodule M
$Inder(A, M)$	set of inner derivations of a (A, A)-bimodule M
$B^1(A, M)$	set of 1-coboundaries of a (A, A)-bimodule M
$H^1(A, M)$	first Hochschild cohomology group of a (A, A)-bimodule M
$Z^2(A, M)$	set of 2-cocycles of a (A, A)-bimodule M
$B^2(A, M)$	set of 2-coboundaries of a (A, A)-bimodule M
$H^2(A, M)$	second Hochschild cohomology group of a (A, A)-bimodule M

Algebras

$(K, A), A^K$	the adjunction of an unit
φ	the embedding-function of A into (K, A)
$\bigoplus_{i=1}^{r} A_i$	the direct sum of K-algebras A_i
A^-, A^{op}	the opposite or inverse algebra of A
$\cdot_{op}$	$a \cdot_{op} b := ba$
A_L	scalar extension of A, $A \otimes_K L$
$Aut_K(A)$	the set of all K-algebra automorphism of a K-algebra A
$Z(A)$	the center of a K-algebra A
$C_A(T)$	the centralizer of the set T of an algebra A
$N_A(T)$	the normalizer of the set T of an algebra A
$\langle T \rangle_{\trianglelefteq A}$	the ideal-span of the subset T within an algebra
α_a	the shift of a within an algebra
$N(A)$	the set of all divisors of zero within an algebra A
$J(A)$	the Jacobson radical of an algebra A

Associative algebras

D_n	the Solomon algebra
$\Delta u, n$	the set of lower triangular matrices of $K^{n \times n}$
$\Delta o, n$	the set of upper triangular matrices of $K^{n \times n}$
$s\Delta u, n$	the set of strict lower triangular matrices of $K^{n \times n}$
$s\Delta o, n$	the set of strict upper triangular matrices of $K^{n \times n}$
$D(n, K)$	the set of diagonal matrices of $K^{n \times n}$
$E(A)$	the group of units of an associative unitary algebra A
κ_e	the conjugation by an unit e within an associative algebra
a^e resp. T^e	$a\kappa_e$ resp. $T\kappa_e$
$*$	the star or circle or quasi regular composition
$Q(A)$	the quasi regular or star or circle group of an associative algebra A
A^*	another notation for $Q(A)$
$e^{(-1)}, e'$	the inverse of a quasi regular element e
$\kappa_{(e)}$	the conjugation with a quasi regular element e
$a^{(e)}$ resp. $T^{(e)}$	$a\kappa_{(e)}$ resp. $T\kappa_{(e)}$
$rad(A)$	the nilradical of an associative algebra A
$Nil(A)$	the set of nilpotent elements of an associative algebra A
KG	the group algebra based on a group G and a field K
$Aug(KG)$	the augmentation ideal of KG
$cl(A)$	the class of nilpotency of an associative algebra A
$cl(a)$	the class of nilpotency of an element a of an associative algebra
ρ	the right regular representation of an associative algebra
λ	the left regular representation of an associative algebra

$\mathcal{R}_1$	the class of unitary rings
$\mathcal{A}$	the class of associative algebras
$\mathcal{A}_1$	the class of associative unitary algebras
$\mathcal{A}$-isomorphic, $\cong_{\mathcal{A}}$	isomorphism within the class $\mathcal{A}$
$\mathcal{A}_1$-isomorphic, $\cong_{\mathcal{A}_1}$	isomorphism within the class $\mathcal{A}_1$
$\langle T \rangle_{\mathcal{A}_1}$	the algebra span of T within $\mathcal{A}_1$
$\langle T \rangle_{\mathcal{A}}$	the algebra span of T within $\mathcal{A}$
$A^{<n>}$	the n-th power of an associative algebra A
$<,>_\rho$	the standard trace form of ρ
$<,>_\lambda$	the standard trace form of λ
$<a,b>_{\lambda,\rho}$	$= tr(a\lambda\, b\rho + a\rho\, b\lambda)$
$A(a,b,K), A(a,b)$	the generalized quaternion algebra
$S(A_i, n)$	see image 8
Z_A	see image 8
$A^{n \times m}$	the set of $n \times m$-matrices over A

Lie algebras

A°	the associated Lie algebra
$\circ$	$a \circ b := ab - ba$
$L^{(n)}$	the n-th derived K-subalgebra of a K-Lie algebra L
$cl(L)$	the nilpotency class of a nilpotent K-Lie algebra L
$S \circ T$	the K-linear span of the set $\{s \circ t \mid s \in S, t \in T\}$
ad	the adjoint representation
$st(L)$	the solvable class of a solvable K-Lie algebra L

Solvable algebras

$AUF(A)$	the solvable radical of an associative K-algebra A
$auf(A)$	the solvable residuum of an associative K-algebra A
A'	the derived K-subalgebra of a K-algebra A
$A^{(n)}$	the n-th derived K-subalgebra K-algebra A
$st(A)$	the solvable class of a solvable K-algebra A

Commutative algebras

$H(A)$	the set of semisimple elements of an algebra A
$D(A)$	the set of diagonalizable elements of an algebra A
$Sep(A)$	the set of separable elements of an algebra A
$VSep(A)$	the set of fully separable elements of an algebra A
$ZF(A)$	the set of splitting elements of an algebra A
$min_{a,K}, \widetilde{min}_{a,K}$	the minimum polynomial of a over a field K
F_a	isomorphism between $K[a]$ and $K[t]/(min_{a,K})$

Notation

$\widetilde{F}_a$	the inserting of a into $tK[t]$
χ	the isomorphisms within the Chinese Remainder theorem
$char_{a,K}$	the characteristical polynomial of a over a field K
aug	the augmentation function of KG onto K
χ_i	an irreducible character of a group
e_i	an idempotent related to χ_i
ω_d	a primitive d-th root of unity
ϕ	the Phi-function
$Gal(L;K)$	the Galois group of a field extension $(K;L)$
$h(G)$	the class number of a group G
$K(\omega_d)$	the adjunction of a primitive d-th root of unity to the field K
$Irr_K(G)$	the set of all irreducible characters of a group G.

Chapter 1

Separable algebras and the theorem of Wedderburn-Malcev

1.1 Separable algebras

1.1.1 Characterizations, properties and examples

Within this section we provide a short introduction to separable algebras. The reader may also read and study the corresponding chapters within the text books of Richard Pierce [35] and Yurij Drozd [8].

Definitions 1 For all $n \in \mathbb{N}$ we define $\underline{n} := \mathbb{N}_{\leq n}$. The symbols $\mathcal{R}_1$, $\mathcal{A}$ resp. $\mathcal{A}_1$ are used for the classes of unitary rings, associative algebras resp. associative unitary algebras. If $\mathcal{K}$ is one the classes $\mathcal{R}_1$, $\mathcal{A}$ or $\mathcal{A}_1$, then $\langle ... \rangle_{\mathcal{K}}$ resp. $\cong_{\mathcal{K}}$ denotes the span resp. the isomorphism within the class $\mathcal{K}$. In addition, we use the word $\mathcal{K}$-isomorphism or say that two objects of the class $\mathcal{K}$ are $\mathcal{K}$-isomorphic. By A^- resp. A^{op} we denote the opposite algebra of an algebra A with respect to the multiplication $a \cdot_{op} b := ba$ for all $a, b \in A$. The center of an algebra A is symbolized by $Z(A)$.⋄

Within this work we use definitions used within module theory like module, algebra module, (A, B)-bimodule, semisimple module, projective module, irreducible module etc. An (A, A)-bimodule is also called A-bimodule within this work. The reader may also read and study the relevant chapters within the text books of Richard Pierce [35] and Yurij Drozd [8]. From algebra and linear algebra theory we use terms K-algebra, K-subalgebra, K-ideal, K-right ideal, K-left ideal, K-space, K-subspace etc. If the connection to the field K is unambiguous, then we omit it and use the terms algebra, subalgebra etc.

Definition 1 *(separable algebra)* An associative unitary K-algebra A is called separable if and only if A is projective as $A^- \otimes_K A$-algebra module (see [35], section 10.2, definition). The next characterization shows us how to detect this property inside the algebra itself. In addition, separable algebras are closely connected to separable field extension.◊

Theorem 1 *(characterizations of separable algebras)* Let K be a field and A an associative unitary K-algebra. The following statements are equivalent:

(i) A is separable.

(ii) $A \otimes A^-$ possesses a so-called separating idempotent: an element $t \in A \otimes A^-$ exists such that $\mu(t) = 1_A$ and $at = ta$ for every $a \in A$ are valid. Here μ is related to the multiplication of A (μ is defined by $\mu(a \otimes b) := ab$) and the expression $at = ta$ is noted within the A-bimodule structure of $A \otimes A^-$.

(iii) $A^- \otimes_K A$ is semisimple and finite-dimensional.

(iv) For every field extension $(K; L)$ the L-algebra $A_L := A \otimes_K L$ (basic field or scalar extension) is semisimple.

(v) A natural number $r \in \mathbb{N}$ and associative finite-dimensional unitary simple K-algebras $A_1, ..., A_r$ exist such that $A \cong_{A_1} \bigoplus_{i=1}^{r} A_i$ is valid and for every $i \in \underline{r}$ the pair $(K1_{A_i}; Z(A_i))$ is a separable field extension.

Proof. see theorem 6.1.2 in [8] and the proposition on page 182 in [35].◊

Based on this theorem we can deduce some properties and also present some examples of separable algebras.

Corollary 1 *(properties and examples of separable algebras)* Let K be a field and A an associative unitary K-algebra.

(i) If A is separable, then A is semisimple and finite-dimensional.

(ii) If A is separable, then $Z(A)$ is separable.

(iii) Direct products of separable algebras are separable.

(iv) For every $n \in \mathbb{N}$ the K-algebra K^n is separable.

(v) Direct products of full matrix algebras over K are separable.

(vi) Let K be algebraical closed. A is separable if and only if A is finite-dimensional and semisimple.

Separable algebras and the theorem of Wedderburn-Malcev 15

(vii) Let K be perfect. A is separable if and only if A is finite-dimensional and semisimple.

(viii) Let $(K;L)$ be a finite-dimensional field extension. The following statements are valid:

  (a) $(K;L)$ is a separable field extension.

  (b) L is separable as K-algebra.

(ix) A direct product of separable field extensions of K is a separable K-algebra.

Proof. The proof is a direct consequence of part (iv) of theorem 1.◊

Examples 1 (i) $\mathbb{C}$ is – based on part (ii) of corollary 1 – a separable $\mathbb{R}$-algebra.

(ii) $\mathbb{R}$ is not finite-dimensional as $\mathbb{Q}$-algebra, and hence – based on part (i) of corollary 1 – not separable as $\mathbb{Q}$-algebra.

(iii) Let K be a field and A an associative n-dimensional unitary central-simple K-algebra. A is – based on part (iv) of corollary 1 – separable. In particular, the quaternion algebra $\mathbb{H}$ is separable as $\mathbb{R}$-algebra and for all $n \in \mathbb{N}$ the K-algebra $K^{n \times n}$ is separable.◊

1.1.2 Group algebras and separability

Within this section we analyze on what terms the group algebra is separable. Let K be a field and G a finite group. By $char(K)$ we denote the characteristic of the field K. The group algebra is symbolized by KG.

Remark 1 Let K be a field and G a finite group. The following statements are valid:

(i) $KG \cong_{\mathcal{A}_1} (KG)^-$

(ii) For every finite group H the statement $K(G \times H) \cong_{\mathcal{A}_1} KG \otimes_K KH$ is valid.

Proof. The K-linear extension of the function

$$G \longrightarrow KG, g \longmapsto g^{-1}$$

is a $\mathcal{A}_1$-isomorphism between KG and $(KG)^-$. In addition, the K-algebras $K(G \times H)$ and $KG \otimes_K KH$ are $\mathcal{A}_1$-isomorphic based on the K-linear extension of the map

$$G \times H \longrightarrow KG \otimes_K KH, (g;h) \longmapsto g \otimes h.$$

Theorem 2 *(separability of group algebras) Let K be a field and G a finite group. The following statements are equivalent:*

(i) KG is separable.

(ii) KG is semisimple.

(iii) $char(K)$ is not a divisor of the order of G.

Proof. The equivalence of (ii) and (iii) is the content of the theorem of Maschke. The implication (i) to (ii) can be proven based on part (i) of corollary 1. We need to prove the implication (ii) to (i). By using remark 1 we deduce $KG \otimes_K (KG)^- \cong_{A_1} K(G \times G)$. $char(K)$ is zero or a prime number. Thus, based on the semisimplicity and the theorem of Maschke $K(G \times G)$ is semisimple, too. By using part (ii) of corollary 1 we finish the proof.⋄

1.1.3 Matrix algebras of separable algebras

Within this section we prove that matrix algebras of separable algebras are separable, too.

Remark 2 *(isomorphism of matrix algebras) Let K be a field, $n, m \in \mathbb{N}$ and A an associative unitary finite-dimensional K-algebra. The following statements are valid:*

(i) $(A^{n \times n})^-$ and $(A^-)^{n \times n}$ are isomorphic.

(ii) $A^{n \times n}$ is isomorphic to $K^{n \times n} \otimes_K A$.

(iii) $K^{n \times n} \otimes_K K^{m \times m}$ and $K^{(nm) \times (nm)}$ are isomorphic.

Proof. The reader may execute the proof within the exercises.⋄

Within the Morita-theory of associative algebras the nilradical is determined for matrix algebras:

Remark 3 *(nilradical of matrix algebras) Let $n \in \mathbb{N}$ and A an associative right artian K-algebra. The nilradical of the matrix algebra $A^{n \times n}$ is exactly $rad(A)^{n \times n}$ (which is the matrix algebra of the nilradical). In particular, A is semisimple if and only if $A^{n \times n}$ is semisimple:*

$$rad(A^{n \times n}) = rad(A)^{n \times n}.$$

Proof. The reader may execute the proof as an exercise.⋄

Theorem 3 *(separability of matrix algebras) Let K be a field, $n \in \mathbb{N}$ and A an associative separable K-algebra. The matrix algebra $A^{n \times n}$ is separable.*

Proof. Based on theorem 1 we have to prove that $(A^{n \times n}) \otimes_K (A^{n \times n})^-$ is semisimple. Based on this statement and remark 2 as well as the commutativity and associativity of the tensor product we deduce:

$$(A^{n \times n}) \otimes_K (A^{n \times n})^- \cong_A$$
$$K^{n \times n} \otimes_K A \otimes_K K^{n \times n} \otimes_K A^- \cong_A$$
$$K^{n^2 \times n^2} \otimes_K (A \otimes_K A^-) \cong_A$$
$$(A \otimes_K A^-)^{n^2 \times n^2}.$$

A is separable, and thus $A \otimes A^-$ is semisimple based on theorem 1. By using remark 3 the matrix algebra $(A \otimes A^-)^{n^2 \times n^2}$ is semisimple and the proof is finished◇

1.1.4 Separable algebras, derivations and factor sets

1.1.4.1 Derivations

We begin this section by defining derivations within the context of bimodules.

Definition and remark 1 *(derivation, inner derivation, first Hochschild cohomology group)* Let A be an associative K-algebra and M an A-bimodule. A K-linear map $d : A \longrightarrow M$ is referred to as a derivation or 1-cocycle from A into M if

$$d(ab) = a.d(b) + d(a).b$$

is valid for all $a, b \in A$. By $Der(A, M) = Z^1(A, M)$ we denote the set of all derivations from A in to M. Given $m \in M$, the linear map

$$ad(m) : A \longrightarrow M, a \mapsto a.m - m.a$$

is the inner derivation or 1-coboundary effected by m. By $Inder(A, M) = B^1(A, M)$ we denote the set of all inner derivations from A into M. If we consider A as A-bimodule the set $Der(A) := Der(A, A)$ resp. $Inder(A) := Der(A, A)$ is the collection of all derivations resp. inner derivations of A. Using the bar resolution one sees that the first Hochschild[1] cohomology group $H^1(A, M)$ is the factor space of derivations by inner derivations from

---
[1] Gerhard Paul Hochschild (born April 29, 1915 in Berlin, died July 8, 2010 in El Cerrito, California) was a German-born American mathematician who worked on Lie groups, algebraic groups, homological algebra and algebraic number theory. Hochschild wrote his thesis in 1941 at Princeton University with Claude Chevalley on Semisimple Algebras and Generalized Derivations. From 1956 up to 1957 he was at the Institute for

A into M. For more details see e.g. [35], [9], [10], [14], [15] and [16]. By $Aut(A)$ or $Aut_K(A)$ resp. $Ant(A)$ or $Ant_K(A)$ we denote the set of all algebra automorphism resp. anti-automorphism of A.⋄

Examples 2 *(examples of derivations)* Let A be an associative K-algebra, M an A-bimodule, $d \in Der(A)$ and $m \in M$.

(i) d_m is a derivation.

(ii) $\mathbb{R}[t]$ possesses no inner derivation different from zero. The formal derivation of polynomials is a derivation.

(iii) Let K be a field and $A := K[t]/(t^2)$. If $char(K) \neq 2$, then $Der(A)$ is of dimension 1 and $Inder(A)$ is of dimension 0.

(iv) Let $\mu : A \otimes_K A \to A$ be the multiplication morphism of the K-algebra A. Consider the A-A-bimodule $A \otimes_K A$. $\ker\mu$ is a sub-bimodule of $A \otimes_K A$ (since μ is an A-A-bimodule map). Consider the map

$$\delta : A \to \ker\mu, \ a \mapsto a \otimes 1 - 1 \otimes a$$

and prove that δ is a derivation.

(v) $Der(A, M)$ is a K-space.

(vi) The kernel of a derivation is a (unital) subalgebra of A.

(vii) Let $d \in Der(A, M)$ and $g \in Aut(A)$. The map $d^g := g^{-1}dg$ is a derivation of A into M.

(viii) Let $d \in Der(A, M)$ and $g \in Ant(A)$. The map $d^g := g^{-1}dg$ is a derivation of A into M.

Proof. ad(i): Let $a, b \in A$. We calculate

$$d_m(ab) = (ab)m - m(ab)$$

and

$$d_m(a)b + ad_m(b) =$$
$$(am - ma)b + a(bm - mb) =$$
$$(am)b - (ma)b + a(bm) - a(mb) =$$
$$d_m(ab).$$

Advanced Study. He was professor at the University of Illinois at Urbana-Champaign and from the end of the 1950s at the University of California, Berkeley. Hochschild introduced Hochschild cohomology, a cohomology theory for algebras, which classifies deformations of algebras. Hochschild and Nakayama introduced cohomology into class field theory. Along with Bertram Kostant and Alex F. T. W. Rosenberg, the Hochschild-Kostant-Rosenberg theorem is named after him. Among his students were Andrzej Bialynicki-Birula and James Ax. In 1955 he was a Guggenheim Fellow. In 1979 he was elected to the National Academy of Sciences, and in 1980 he was awarded the Leroy P. Steele Prize of the AMS.

ad(ii): The algebra is commutative and possesses therefor no inner derivation different from zero. The other statement is valid because of the following well-know rules:

$$(fg)' = f' + g'$$
$$(kf)' = kf'$$
$$(fg)' = f'g + fg'.$$

These are the rules for being a derivation.

ad(iii): Let K be a field. Consider the commutative, unitary, associative and 2-dimensional algebra $A := K[t]/(t^2)$. This algebra possesses a basis $\{1, r\}$ such that $r^2 = 0$ is valid. A is commutative and therefore $Inder(A)$ is the zero space. Let d be a derivation of A. We calculate

$$d(1) = d(1 \cdot 1) = 1 \cdot d(1) + d(1) \cdot 1 = 2 \cdot d(1).$$

Hence, $d(1) = 0$ is true. Let $char(K) \neq 2$. Because of $r^2 = 0$ we derive

$$0 = d(r^2) = 2rd(r).$$

Let $k, l \in K$ such that $d(r) = k1 + lr$ is valid. We calculate

$$0 = 2rd(r) = 2kr.$$

Thus, $k = 0$ is valid. The derivation is defined by $d(1) = 0$ and $d(r) = lr$. If we define a linear function by these rules, then we can prove that this linear function is indeed a derivation. Let $x_1, x_2 \in A$ and $r, s, t, u \in K$ such that $x_1 = n1 + sr$ and $x_2 = t1 + ur$ are valid. A straightforward calculation shows

$$d(x_1 x_2) = l(nu + st)r = d(x_1)x_2 + x_1 d(x_2).$$

Thus, the set of derivations is of dimension 1 possessing no inner derivations different from zero.

ad(iv): Let $a, b \in A$. We calculate

$$(ab)\delta = (ab) \otimes 1 - 1 \otimes (ab).$$

In addition, the following calculation is valid:

$$(a\delta).b + a.(b\delta) = (a \otimes 1 - 1 \otimes a).b + a.(b \otimes 1 - 1 \otimes b).$$

From left, A acts on the left component, and from right A acts on the right components of the tensors within this bimodule structure. Thus, we derive

$$(a\delta).b + a.(b\delta) = (a \otimes b) - (1 \otimes (ab)) + ((ab) \otimes 1) - (a \otimes b) = (ab) \otimes 1 - 1 \otimes (ab).$$

ad(v): Let $d, e \in Der(A, M)$, $a, b \in A$ and $k \in K$. We calculate

$$(ab)(d+e) = (ab)d + (ab)e = a(bd) + (ad)b + a(be) + (ae)b = a(b(d+e)) + (a(d+e))b$$

and

$$(ab)(ke) = k(ab)e = k((ae)b + a(be)) = k(ae)b + ka(be) = a(ke)b + a(b(ke)).$$

ad(vi): The kernel of a K-linear function is a K-space. Let $d \in Der(A, M)$ and $a, b \in ker(d)$. We calculate

$$d(ab) = ad(b) + d(a)b = 0 + 0 = 0.$$

Furthermore, we derive:

$$d(1) = d(1 \cdot 1) = 1 \cdot d(1) + d(1) \cdot 1 = 2 \cdot d(1).$$

Thus, we conclude $d(1) = 0$.

ad(vii): Let $a, b \in A$. We calculate

$$\begin{aligned}
(ab)d^g &= \\
(ab)g^{-1}dg &= \\
(ag^{-1})(bg^{-1})dg &= \\
((ag^{-1})d(bg^{-1}) + (ag^{-1})(bg^{-1})d)g &= \\
(ag^{-1})dg \cdot b + a(ag^{-1})dg. &
\end{aligned}$$

ad(viii): This statement can be derived from part (vii) by using the opposite algebra A^{op} or by direct calculation as done within part (vii).◇

Bi-module derivations are closely connected to separable algebras. This is the content of the next theorem. This characterization is used within the appendix for proving the theorems of Wedderburn-Malcev and Taft.

Theorem 4 *(characterization of separable algebras by inner derivations)* *Let R be a field and A be an unital R-algebra. Then, A is a separable R-algebra if and only if every derivation from A to an A-A-bimodule is inner.*

Proof. (see [63]) $\Longrightarrow$: Assume that A is a separable R-algebra. Then, based on theorem 1, there exists an element $t \in A \otimes_R A$ (where all tensor products are over R) satisfying $\mu(t) = 1$ (where $\mu : A \otimes_R A \to A$ is the multiplication morphism of the R-algebra A) and $at = ta$ for every $a \in A$ (where we are using the standard A-A-bimodule structure on $A \otimes_R A$). Now, let M be an A-A-bimodule, and $d : A \to M$ be a derivation. Since $t \in A \otimes_R A$

is a tensor, we can write it in the form $t = \sum_{i=1}^{n} t_i \otimes s_i$ for some $n \in \mathbb{N}$ and some $t_1, t_2, ..., t_n \in A$ and $s_1, s_2, ..., s_n \in A$. Then, $\mu(t) = \sum_{i=1}^{n} t_i s_i$, so that $\mu(t) = 1$ becomes $\sum_{i=1}^{n} t_i s_i = 1$. On the other hand, every $a \in A$ satisfies $at = ta$. Since $t = \sum_{i=1}^{n} t_i \otimes s_i$, this rewrites as

$$\sum_{i=1}^{n} at_i \otimes s_i = \sum_{i=1}^{n} t_i \otimes s_i a.$$

Applying the map $d \otimes_R \mathrm{id}$ to this equation, we get

$$\sum_{i=1}^{n} d(at_i) \otimes s_i = \sum_{i=1}^{n} d(t_i) \otimes s_i a.$$

Applying the action map

$$A \otimes M \to M, \ a \otimes m \mapsto am$$

to this equation, we get

$$\sum_{i=1}^{n} d(at_i) s_i = \sum_{i=1}^{n} d(t_i) s_i a.$$

Thus,

$$0 = \sum_{i=1}^{n} d(at_i) s_i - \sum_{i=1}^{n} d(t_i) s_i a =$$

$$\sum_{i=1}^{n} \left(\underbrace{d(at_i)}_{=d(a)t_i + ad(t_i) \text{ (since } d \text{ is a derivation)}} s_i - d(t_i) s_i a \right) =$$

$$\sum_{i=1}^{n} \left(d(a) t_i s_i + a d(t_i) s_i - d(t_i) s_i a \right) =$$

$$d(a) \underbrace{\sum_{i=1}^{n} t_i s_i}_{=1} + a \sum_{i=1}^{n} d(t_i) s_i - \sum_{i=1}^{n} d(t_i) s_i a =$$

$$d(a) + a \sum_{i=1}^{n} d(t_i) s_i - \sum_{i=1}^{n} d(t_i) s_i a.$$

Hence,
$$d(a) = -a\sum_{i=1}^{n} d(t_i) s_i + \sum_{i=1}^{n} d(t_i) s_i a.$$

In other words,
$$d(a) = au - ua$$

where $u = -\sum_{i=1}^{n} d(t_i) s_i$. This shows that d is an inner derivation. We have thus proven that every derivation from A into an A-A-bimodule is inner.

$\Longleftarrow$: Assume that every derivation from A into an A-A-bimodule is inner. Let $\mu : A \otimes_R A \to A$ be the multiplication morphism of the R-algebra A. Consider the A-A-bimodule $A \otimes_R A$; then, $\ker \mu$ is a sub-bimodule of $A \otimes_R A$ (since μ is an A-A-bimodule map, as can be easily seen). Consider the map

$$\delta : A \to \ker\mu, \ a \mapsto a \otimes 1 - 1 \otimes a.$$

This map δ is a derivation (see example 2), so it is inner (by the assumption that every derivation from A into an A-A-bimodule is inner). This means that there exists some $u \in \ker\mu$ such that

$$\delta(a) = au - ua$$

is valid for for every $a \in A$. Consider this u. Let $t = 1 \otimes 1 - u$. Then, $\mu(u) = 0$ (since $u \in \ker\mu$) and $\mu(1 \otimes 1) = 1$ yield $\mu(t) = 1$. On the other hand, every $a \in A$ satisfies:

$$at - ta =$$
$$a(1 \otimes 1 - u) - (1 \otimes 1 - u)a =$$
$$\left(\underbrace{a(1 \otimes 1)}_{=a \otimes 1} - \underbrace{(1 \otimes 1)a}_{=1 \otimes a} \right) - \underbrace{(au - ua)}_{=\delta(a)=a\otimes 1 - 1 \otimes a} =$$
$$(a \otimes 1 - 1 \otimes a) - (a \otimes 1 - 1 \otimes a) = 0.$$

Thus there exists an element $t \in A \otimes A$ such that $\mu(t) = 1$ and $at = ta$ for every $a \in A$. This means that the R-algebra A is separable (see theorem 1).$\diamond$

A direct consequence of this theorem for $M = A$ is:

Corollary 2 *(separable algebras and inner derivations) Let K be a field and A a separable associative algebra. A possesses only inner derivations:*

$$Der(A) = Inder(A). \diamond$$

This statement does not characterize separable algebras. Examples of non-separable algebras exist possessing only inner derivations. For this, we provide two examples. The first one will be just stated, the second one will be discussed in details. The first one is related to the so-called Solomon algebra in characteristic zero which is not semisimple. The dimension of the factor algebra by the nilradical is related to the partition numbers. Within his dissertation [3] T. Bauer proves the following theorem:

Theorem 5 *(derivations of Solomon algebras) Let K be a field of characteristic zero and D_n ($n \in \mathbb{N}$) be a Solomon algebra over K. D_n possess only inner derivations.* $\diamond$

The second example is related to the algebra of upper triangular matrices of $K^{n \times n}$ over an arbitrary field K. Within [5] it is proven that this algebra contains only inner derivations. The set of diagonal matrices is a radical complement, the algebra is not separable. We want to analyze this result based on techniques used by T. Bauer within his dissertation [3]. His idea is first to analyze the restriction of a derivation to a radical complement and prove that the restriction is a restriction of an inner derivation. Afterwards derivations are analyzed which are zero restricted to a radical complement. For these special derivations, he analyzed the so-called Pierce components of the algebra. The Pierce components are invariant under these special derivations and of low dimension. Furthermore, the values of this derivations on all Pierce components can be calculated based on some special Pierce components. By comparing dimensions of these special derivations (which can be bounded by the sum of dimensions of the special Pierce components) with the dimension of the factor algebra of the radical complement by the center of the entire algebra he proves that also these special derivations are inner. Both results ensure that all derivations are inner. We generalize the first step within his argumentation to a wider class of algebras. The proof is essentially the same as stated within [3].

Lemma 1 *Let K be a field, A an unitary associative finite-dimensional K-algebra, $n \in \mathbb{N}$ such that $A/rad(A)$ is isomorphic to K^n, T a radical complement in A and d a derivation of A in A. The following statements are valid:*

(i) An element $x \in A$ exists such that $d_{|T} = ad(x)_{|T}$.

(ii) Let $d_{|T} = 0$ and $e_1, \ldots, e_n$ the primitive orthogonal idempotents of T. For all $i, j \in \mathbb{N}$ the subspace $e_i A e_j$ is a d-invariant Pierce-component of A.

Proof. ad(i): Based on remark 4 we derive

$$1_T = 1_A = \sum_{i=1}^{n} e_i.$$

Let $i, j \in \mathbb{N}$ such that $i \neq j$ is valid. Then, $e_i e_j = 0$ is valid because the idempotents are orthogonal. We conclude

$$0 = d(0) = d(e_i e_j) = (e_i d)e_j + e_i(e_j d).$$

We define $x = \sum_{i=1}^{n} e_i(e_i d)$ and calculate for an arbitrary $k \in \mathbb{N}$:

$$e_k ad(x) =$$
$$e_k \circ x =$$
$$e_k \circ \sum_{i=1}^{n} e_i(e_i d) =$$
$$e_k \sum_{i=1}^{n} e_i(e_i d) - (\sum_{i=1}^{n} e_i(e_i d))e_k.$$

Because of $e_i e_j = 0$ and $(e_k)^2 = e_k$ we derive

$$e_k ad(x) =$$
$$e_k(e_k d) - \sum_{i=1}^{n} e_i(e_i d)e_k =$$
$$e_k(e_k d) - \sum_{i=1}^{n} e_i(e_i d)e_k e_k.$$

Thus, we conclude by using $0 = (e_i d)e_j + e_i(e_j d)$ and $(e_i)^2 = e_i$:

$$e_k ad(x) =$$
$$e_k(e_k d) + \sum_{i=1}^{n} e_i e_i(e_k d)e_k =$$
$$e_k(e_k d) + \sum_{i=1}^{n} e_i(e_k d)e_k =$$
$$e_k(e_k d) + (\sum_{i=1}^{n} e_i)(e_k d)e_k =$$
$$e_k(e_k d) + (e_k d)e_k 1_A.$$

e_k is idempotent, and thus

$$e_k d = ((e_k)^2)d = (e_k d)e_k + e_k(e_k d)$$

is valid. We finalize now the proof of part (i):

$$e_k ad(x) = e_k(e_k d) + (e_k d)e_k 1_A = e_k(e_k d) + (e_k d)e_k = e_k d.$$

ad(ii): Let $a \in A$. We calculate

$$\begin{aligned}
(e_i a e_j)d &= \\
(e_i d) a e_j + e_i(a e_j)d &= \\
0 + e_i(ad)e_j + e_i a(e_j d) &= \\
e_i(ad)e_j + 0 &= \\
e_i(ad)e_j &\in e_i A e_j.
\end{aligned}$$

Thus, the lemma is proven.◇

We proceed by stating some facts about the algebra of upper triangular matrices over a field. (The reader may prove these statements in details within exercise 29.) Let K be a field and $n \in \mathbb{N}$. By $\Delta_{u,n}$ resp. $\Delta_{o,n}$ we denote the set of all lower resp. upper triangular matrices of $K^{n \times n}$. Both set are subalgebras of the K-algebra $K^{n \times n}$. In addition, $rad(\Delta_{u,n})$ resp. $rad(\Delta_{o,n})$ is the set of all strict lower resp. strict upper triangular matrices of $K^{n \times n}$. In both cases the factor algebra of its nilradical is $\mathcal{A}_1$-isomorphic to the separable K-algebra K^n (see part (iv) of theorem 1). In addition, $D(n, K)$, the set of all diagonal matrices of $K^{n \times n}$ is a radical complement for both algebras. The structure is visualized by the following Hasse diagram:

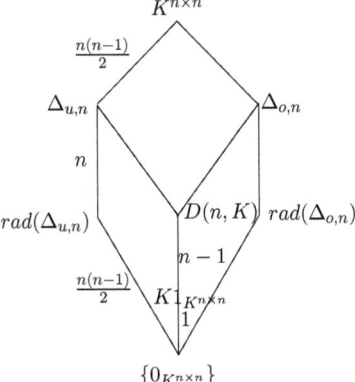

For every $i, j \in \mathbb{N}$ let $E_{i,j}$ be that matrix possessing only one entry different to zero which is 1 in the i-j-cell. The set of all matrices $\{E_{i,j} \mid i, j \in \underline{n}, i \leq j\}$ is a K-basis of $\Delta_{o,n}$. For every $i, j, r, s \in \underline{n}$ the identities $E_{i,j}E_{r,s} = E_{i,s}$ for $j = r$ and $E_{i,j}E_{r,s} = 0$ for $j \neq r$ are valid.

Proposition 1 Let K be a field and $n \in \mathbb{N}$. The following statements are valid:

(i) For all $i,j \in \underline{n}$ such that $i \leq j$ the identity $E_{ii}\Delta_{o,n}E_{j,j} = \langle E_{i,j}\rangle_K$ is valid.

(ii) $\Delta_{o,n}$ is central: $Z(\Delta_{o,n}) = K \cdot 1_{\Delta_{o,n}}$.

(iii) For all $i,j \in \underline{n}$ such that $i < j$ the identity $E_{i,j} = E_{i,i+1}E_{i+1,i+2}\cdots E_{j-1,j}$ is valid. The elements on the second diagonal of $\Delta_{o,n}$ are generating the set of strict upper triangular matrices as an algebra.

Proof. ad(i): Let $i,j \in \underline{n}$ such that $i \leq j$ is valid. We calculate – by using the multiplication rule for the basis elements –

$$E_{i,i}\Delta_{o,n} = E_{i,i}\bigoplus_{r \leq k}\langle E_{r,k}\rangle_K = \bigoplus_{i \leq k}\langle E_{i,k}\rangle_K.$$

Thus,

$$E_{i,i}\Delta_{o,n}E_{j,j} = \bigoplus_{i \leq k}\langle E_{i,k}\rangle_K E_{j,j} = \langle E_{i,j}\rangle_K$$

is true.

ad(ii): By using part (i) and lemma 5 on page 81 in [52] we derive that all Pierce components $E_{i,i}\Delta_{o,n}E_{i,i}$ are one-dimensional, and thus the radical complement $D(n,K) = \langle E_{1,1},\ldots,E_{n,n}\rangle_K$ is self-centralizing in $\Delta_{o,n}$. We conclude, that the center of $\Delta_{o,n}$ is contained in $D(n,K)$. If a diagonal matrix is commuting with all elements of $\Delta_{o,n}$, then its transpose is commuting with all elements of the transpose $(\Delta_{o,n})^T$. The elements of $D(n,K)$ are fix points under ^{T}, and thus the center of $\Delta_{o,n}$ is contained in the center of $K^{n\times n}$ which is a central algebra.

ad(iii): This identity is a direct consequence of the multiplication rule of the basis elements $E_{r,s}$. ⋄

Now we can prove that $\Delta_{o,n}$ possesses only inner derivation:

Theorem 6 *(derivations of the algebra of upper triangular matrices)* Let K be a field and $n \in \mathbb{N}$. $\Delta_{o,n}$ possesses only inner derivations:

$$Der(\Delta_{o,n}) = Inder(\Delta_{o,n}).$$

Proof. Let $A := \Delta_{o,n}$ and $d \in Der(A)$. We use lemma 1 and derive that an element $x \in A$ exists such that $d_{|D(n,K)} = ad(x)_{|D(n,K)}$ is valid. Let $f := d - ad(x)$. Then f is a derivation of A such that $f_{|D(n,K)} = 0$. We

Separable algebras and the theorem of Wedderburn-Malcev

want to prove that f is an inner derivation of A which finishes the proof. If $y \in A$ exists such that $f = ad(y)$ is valid, then $d - ad(x) = ad(y)$ and hence $d = ad(y) + ad(x) = ad(x + y)$ are valid. Again by using lemma 1 the Pierce components $E_{i,i}AE_{j,j}$ are f-invariant for all $i, j \in \underline{n}$ such that $i \leq j$ is valid. In particular, the Pierce components $E_{i,i}AE_{i+1,i+1}$ are d-invariant for all $i \in \underline{n-1}$. In view of proposition 1 all of the components are one-dimensional. Hence, for f we have $n - 1$-dimensional 'choices'. All other values of f are determined by these special Pierce components based on part (iii) of proposition 1 and the derivation rule. We conclude (by using also part (ii) of proposition 1) that

$$n - 1 =$$
$$dim(D(n, K)/Z(A)) =$$
$$dim\{g \mid g = ad(t), t \in D(n, K)\} \leq$$
$$dim\{g \mid g \in Der(A), g_{|D(n,K)} = 0\} \leq$$
$$n - 1$$

is valid and that f is a inner derivation of A.⋄

1.1.4.2 Factor sets

We begin this section by defining factor sets within the context of bimodules.

Definition and remark 2 *(factor sets, splitting factor sets, second Hochschild cohomology group)* Let A be an associative K-algebra and M an A-bimodule. A K-bilinear map $f : A \times A \longrightarrow M$ is referred to as a factor set or 2-cocycle from A into M if

$$a.f(b,c) - f(ab,c) + f(a,bc) - f(a,b).c = 0$$

for all $a, b, c \in A$ is valid. By $Z^2(A, M)$ we denote the set of all factor sets from A into M. If there exists a K-linear map $g : A \longrightarrow M$ with

$$f(a,b) = a.g(b) - g(ab) + g(a).b$$

for all $a, b \in A$, then f is called a split factor set or 2-coboundary from A into M. By $B^2(A, M)$ we denote the collection of all split factor sets from A into M. Using the bar resolution one sees that the second Hochschild cohomology group $H^2(A, M)$ is the factor space of the factor sets by the split factor sets of A into M. For more details see e.g. [35], [9], [10], [14], [15] and [16].⋄

Separable algebras possess a trivial second Hochschild cohomology group which is the context of the next theorem. This result is used within the appendix for proving the theorem of Wedderburn-Malcev.

Theorem 7 *(factors sets of separable algebras) Let A be an associative R-algebra and M an A-bimodule. If A is separable, then $B^2(A, M)$ is trivial.*

Proof. The proof is similar as done within theorem 4 for proving that separable algebras possess only inner derivations and is based on the proof presented in [6]. Assume that A is a separable R-algebra. Then, based on theorem 1, there exists an element $t \in A \otimes_R A$ (where all tensor products are over R) satisfying $\mu(t) = 1$ (where $\mu : A \otimes_R A \to A$ is the multiplication morphism of the R-algebra A) and $a.t = t.a$ for every $a \in A$ (where we are using the standard A-A-bimodule structure on $A \otimes_R A$). Now, let M be an A-A-bimodule. Since $t \in A \otimes A$ is a tensor, we can write it in the form $t = \sum_{i=1}^{n} t_i \otimes s_i$ for some $n \in \mathbb{N}$ and some $t_1, t_2, ..., t_n \in A$ and $s_1, s_2, ..., s_n \in A$. Then, $\mu(t) = \sum_{i=1}^{n} t_i s_i$, so that $\mu(t) = 1$ becomes $\sum_{i=1}^{n} t_i s_i = 1$. On the other hand, every $b \in A$ satisfies $bt = ba$. Since $t = \sum_{i=1}^{n} t_i \otimes s_i$, this rewrites as

$$\sum_{i=1}^{n} bt_i \otimes s_i = \sum_{i=1}^{n} t_i \otimes s_i b.$$

Let f be a factor set of A into M. We define

$$g : A \longrightarrow M, a \mapsto \sum_{i=1}^{n} f(a, t_i) s_i.$$

Let $a, b \in A$. We calculate

$$aF(b) - F(ab) + F(a)b = \sum_{i=1}^{n} af(a, t_i)s_i - \sum_{i=1}^{n} f(ab, t_i)s_i + \sum_{i=1}^{n} f(a, t_i)s_i b.$$

Using the factor set rule on the first summand we derive

$$aF(b) - F(ab) + F(a)b = \sum_{i=1}^{n} f(a,b)s_i t_i + \sum_{i=1}^{n} f(a, s_i)t_i b - \sum_{i=1}^{n} f(a, bs_i)t_i.$$

Because of $\mu(t) = 1$ we derive:

$$aF(b) - F(ab) + F(a)b = f(a,b) + \sum_{i=1}^{n} f(a, s_i)t_i b - \sum_{i=1}^{n} f(a, bs_i)t_i.$$

Thus, we need to show that $\sum_{i=1}^{n} f(a, s_i)t_i b - \sum_{i=1}^{n} f(a, bs_i)t_i$ is equal to zero. For proving this we use the rule $b.t = t.b$ and apply the function $f(a; \cdot) \otimes id$ on $b.t = t.b$ where $f(a; \cdot)(x) := f(a; x)$ for all $x \in A$. The proof is finished using the fact that f is bilinear.⋄

1.2 Radical complements and the theorem of Wedderburn-Malcev

For presenting the theorem of Wedderburn-Malcev (see [31]) we need the following definitions.

Definitions 2 *(radical complement)* Let A be a K-algebra and U a K-subspace of A. A K-subalgebra T of A is called an algebra complement of U in A if and only if T a K-space complement of U in A. If A is associative and $U = rad(A)$ the nilradical of A, then T is called a radical complement in A. Analogue we define left ideal, right ideal and ideal complements of K-subspaces of A.⋄

Definition and remark 3 *(conjugation)* Let A be an associative unitary K-algebra. By $E(A)$ we denote the group of units of A. For all $e \in E(A)$ we define

$$\kappa_e : A \longrightarrow A, x \longmapsto x^e := e^{-1}xe.$$

In addition, let

$$\kappa : E(A) \longrightarrow Aut_K(A), e \longmapsto \kappa_e.$$

For all $e \in E(A), a \in A$ and $T \subseteq A$ we use a^e resp. T^e instead of $a\kappa_e$ resp. $T\kappa_e$. If $e \in E(A)$, then the function κ_e is called the conjugation with e.⋄

Theorem 8 *(theorem of Wedderburn-Malcev for unitary algebras)* Let K be a field and A a finite-dimensional associative unitary K-algebra. If $A/rad(A)$ is separable, then the following statements are valid:

(i) A possesses a radical complement.

(ii) The nilpotent normal subgroup $1_A + rad(A)$ of $E(A)$ acts transitive per conjugation on the set of all radical complements of A.

Proof. The original proofs are included in the papers of Wedderburn [49] and Malcev [31]. Hochschild uses cohomology theory of associative algebras to prove the Wedderburn-Malcev theorem. This is presented in his papers [14], [15] and [16]. Within the textbook of Pierce [35] and Curtis/Reiner [6] Hochschild's approach is used, too. Also Farnsteiner presented this approach within his papers [9] and [10]. Within the textbook of Drozd and Kirichenko an approach is given based on a system of linear equation. A similar concept is used within the paper [13] of de Graaf.⋄

We present a proof based on the mentioned literature within the appendix (see chapter B). The methods of de Graaf and Drozd and Kirichenko can be used to define algorithm for calculating a radical complement. This is presented in chapter 2 of this work.

Note 1 The theorem of Wedderburn-Malcev is often applied within the following context: If K is a perfect field and A an associative finite-dimensional unitary K-algebra, then the assumptions of the theorem 8 are met by using part (ii) of corollary 1. Other applications are possible is if the factor algebra by the nilradical is presentable as within parts (iii)-(v) of the same corollary or if it is isomorphic to one of the separable algebras mentioned within this section.◇

1.3 Examples within the context of the theorem of Wedderburn-Malcev

Within this subsection we present examples for the theorem of Wedderburn-Malcev. For this, we start with an example from the exercises of chapter 11.6 in [35]. The mentioned algebra does not possess a radical complement. An algebra possessing two non-conjugated radical complements will be presented later in chapter 4 of this work (about generalized quaternion algebras). Our second examples (again taken from chapter 11.6 of the work [35]) shows us how the theorem of Wedderburn-Malcev can be used. This example is generalized afterwards by analyzing the algebras of lower and upper triangular matrices over a field. For both algebras we calculate all complements explicitly based on a recursion formula. Finally, we give a brief overview of those algebras which arise in the article [38] of Ian Stewart and Martin Golubitsky about dynamical networks. The relevant algebras within their work are studied by using the theorem of Wedderburn-Malcev.

1.3.1 A counterexample and an example

Remark 4 *(radical complements are unital) If A is an associative unitary K-algebra and C is a radical complement of A, then C is an unital K-subalgebra.*

Proof. By using $A = rad(A) \oplus_K C$ exactly one pair $(c; r) \in C \times rad(A)$ exists such that $1_A = c + r$ is valid. We deduce

$$c + r = 1_A = 1_A{}^2 = c^2 + cr + rc + r^2.$$

$rad(A)$ is a K-ideal of A, and thus $c = c^2$ and $r = cr + rc + r^2$ are valid. Because of

$$r = 1_A r = (c + r)r = cr + r^2$$

we deduce $rc = 0_A$, and by using

$$r = r 1_A = r(c + r) = rc + r^2$$

we deduce $cr = 0_A$. We conclude that $r^2 = r$ is valid. Hence, r is nilpotent and idempotent in A. This is only valid for the zero element.◇

Separable algebras and the theorem of Wedderburn-Malcev 31

Definition 2 Let $n \in \mathbb{N}$ and K a field. We define $Pot(n, K) := \{x \mid \exists k \in K : x = k^n\}$ and $QA(K) := Pot(2, K)$. We call $Pot(n, K)$ the set of n-th powers and $QA(K)$ the set of squares of K. ◇

Counterexample 1 *(an algebra possessing no radical complement)* Let K be a field such that $char(K) = 2$ is valid, $F := K(t)$ the field of fractions of the polynomial algebra $K[t]$ in one variable t and A the 4-dimensional unitary F-algebra possessing the F-Basis $C := \{1_A, d, y, z\}$ with respect to the multiplication

·	1_A	d	y	z
1_A	1_A	d	y	z
d	d	$t1_A + y + z$	z	ty
y	y	z	0_A	0_A
z	z	ty	0_A	0_A.

It is straightforward to prove (by using the multiplication table of the basis C) that A is associative and commutative.

Now we prove that $rad(A) = \langle y, z \rangle_F$ is valid.
The multiplication table lets us deduce that $\langle y, z \rangle_F$ is an F-ideal of A. In addition, for all $f_1, f_2 \in F$ the identity

$$(f_1 y + f_2 z)^2 = f_1^2 y^2 + f_1 f_2 yz + f_2 f_1 zy + f_2^2 z^2 = 0_A$$

is true. Hence, $\langle y, z \rangle_F$ is a nil and therefor also a nilpotent F-ideal of A. We have to prove that $A/\langle y, z \rangle_F$ is semisimple. Let $B := A/\langle y, z \rangle_F$. Let us assume that B is not semisimple. 1_A is not nilpotent, and thus B would possess an one-dimensional nilradical. Hence, $rad(B)$ would be a zero-ideal of B. Therefor an element $r \in rad(B)$ would exist such that $r \neq 0_B$ and $r^2 = 0_B$ are valid. Let $h_1, h_2 \in F$ such that $r = (h_1 1_A + h_2 d) + \langle y, z \rangle_F$ is true. By using $r^2 = 0_B$ we would deduce $h_1^2 + h_2^2 t = 0_K$. If $h_2 \neq 0_K$, then $t \in QA(K(t))$ would be valid which is a contradiction. Hence, $h_2 = 0_K$ and $h_1 = 0_K$ are true. We would deduce $r = 0_B$ which is a contradiction. We conclude $rad(A) = \langle y, z \rangle_F$.

Now we prove that A does not possess a radical complement.
We assume a radical complement C of $rad(A)$ in A would exist. By using remark 4 the statement $1_A \in C$ is valid. Let $\{1_A, c\}$ be a F-basis of C, und let $f_1, f_2, f_3, f_4 \in F$ such that $c = f_1 1_A + f_2 d + f_3 y + f_4 z$ is true. By using $c^2 = (f_1^2 + f_2^2 t) 1_A + f_2^2(y + z) \in C$ we deduce $f_2^2(y + z) \in C \cap rad(A) = \{0_A\}$. Hence, $f_2 = 0_K$ is valid. Therefor $f_1 1_A + f_3 y + f_4 z \in C$ and also $f_3 y + f_4 z \in C$ are valid. We deduce $f_3 y + f_4 z \in C \cap rad(A) = \{0_A\}$, and thus $f_3 = f_4 = 0_K$ is valid. We have proven that 1_A and c are not F-linear independent which is a contradiction.

An important reason for this non-existence is that the factor algebra by the nilradical of A is not separable. For proving this fact we show the existence of a non-zero nilpotent element within the commutative F-algebra $T := A/rad(A) \otimes_F A/rad(A)$ (see part (ii) of theorem 1):

$$\begin{aligned}
(((d+rad(A)) \otimes (1_A+rad(A)) + (1_A+rad(A)) \otimes (d+rad(A)))^2 &= \\
(d^2+rad(A)) \otimes (1_A+rad(A)) + (d+rad(A)) \otimes (d+rad(A)) &+ \\
(d+rad(A)) \otimes (d+rad(A)) + (1_A+rad(A)) \otimes (d^2+rad(A)) &= \\
2t((1_A+rad(A)) \otimes (1_A+rad(A))) &= \\
0_T.&
\end{aligned}$$

A two-dimensional unitary F-algebra is not semisimple, a field or $\mathcal{A}_1$-isomorphic to $F \times F$. Thus, the results proven so far and part (iv) of theorem 1 let us deduce that $A/rad(A)$ is an inseparable field extension of $F(1_A + rad(A))$. Indeed, by a straightforward calculation an element $f \in F \setminus QA(F)$ exist such that $t^2 - f$ is the minimal polynomial of $d + rad(A)$ over F and $A/rad(A) = F(1_A + rad(A))[d + rad(A)]$ is valid.⋄

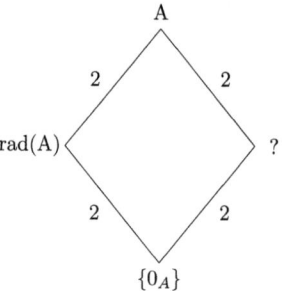

Example 1 Let F be a field and A the three-dimensional unitary F-algebra possessing the F-basis $\{1_A, e, r\}$ with respect to the multiplication table

·	1_A	e	r
1_A	1_A	e	r
e	e	e	r
r	r	0_A	0_A

It is straightforward to prove that A is associative. By using $er = r \neq 0_A = re$ we deduce that A is not commutative.

We prove that $rad(A) = \langle r \rangle_F$ is valid and that $A/rad(A)$ is separable. By using the multiplication table it is straightforward to prove that $\langle r \rangle_F$ is a K-ideal. Because of $r^2 = 0_A$ we deduce that $\langle r \rangle_F$ is a nilpotent F-ideal of A. Now we analyze the factor algebra by this F-ideal of A. The set $\{1_A + \langle r \rangle_F, e + \langle r \rangle_F\}$ is a F-basis of $A/\langle r \rangle_F$. The F-linear extension of the function

$$A/\langle r \rangle_F \longrightarrow F^2, \; 1_A + \langle r \rangle_F \longmapsto (1_F, 1_F), e + \langle r \rangle_F \longmapsto (0_K, 1_K)$$

is a F-algebra isomorphism, and thus $A/rad(A) \cong_{A_1} F^2$ is valid. By using part (iv) of theorem 1 we deduce that $A/\langle r \rangle_F$ is separable. In particular, based on part (i) of corollary 1 the statement $rad(A) = \langle r \rangle_F$ is valid and $A/rad(A)$ is separable.

The Wedderburn-Malcev theorem 8 lets us deduce that A possesses a radical complement C and that all radical complements are of the form $C^{1_A + fr}$ ($f \in F$). It is straightforward to prove that $C := \langle 1_A, e \rangle_F$ is a radical complement of A. Now we determine all radical complements of A. For all $f \in F$ we calculate

$$(1_A + fr)(1_A - fr) = (1_A - fr)(1_A + fr) = 1_A$$

and

$$(1_A - fr)e(1_A + fr) = (e - fre)(1_A + fr) = e + fer - fre - f^2 rer = e + fr.$$

This calculation implies $C^{1_A + fr} = \langle 1_A, e + fr \rangle_F$.

We finish this example by remarking that the function

$$f \longmapsto \langle 1_A, e + fr \rangle_F$$

is a bijection between F and the set of all radical complements of A: we have proven that the function is surjective. Let $f, f' \in F$ such that $\langle 1_A, e + fr \rangle_F = \langle 1_A, e + f'r \rangle_F$ is valid. Elements $k, l \in F$ exists possessing the property $e + fr = k1_A + le + lf'r$. We conclude $k = 0_K$, $l = 1_K$ and $f = lf'$. In particular, $f = f'$ is valid. This statement will be analyzed again within chapter 2.⋄

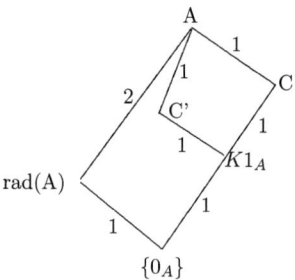

The next section generalizes this example.

1.3.2 The algebras of upper and lower triangular matrices

It is straightforward to present an element of the K-algebra $\Delta_{u,n}$ of lower triangular resp. $\Delta_{o,n}$ of upper triangular matrices into a sum of a strict lower resp. strict upper triangular matrix and a diagonal matrix. We want to calculate all radical complements. Remark 5 implies that we have to calculate this set only for one the algebras $\Delta_{u,n}$ and $\Delta_{o,n}$ (because they are isomorphic).

Definition 3 *(transpose function)* The transpose function $K^{n\times n}$ is defined by
$$\tau : K^{n\times n} \longrightarrow K^{n\times n}, A \longmapsto A^t.\diamond$$

Remark 5 Let $A := \begin{pmatrix} 0_K & \cdot & \cdot & \cdot & 0_K & 1_K \\ 0_K & \cdot & \cdot & \cdot & 1_K & 0_K \\ \cdot & \cdot & \cdot & \cdot & \cdot & \cdot \\ 0_K & 1_K & \cdot & \cdot & 0_K & 0_K \\ 1_K & 0_K & \cdot & \cdot & 0_K & 0_K \end{pmatrix} \in K^{n\times n}$.

The following statements are valid:

(i) A is self-inverse and symmetric. κ_A is a $\mathcal{A}_1$-automorphism of $K^{n\times n}$ of order two. (The image of a matrix M under κ_A is determinable by a point reflection of the values of M at the center of M.)

(ii) $\kappa_A \tau = \tau \kappa_A$

(iii) $\Delta_{u,n} \tau = \Delta_{o,n}$
  In particular, the K-algebras $\Delta_{u,n}$ and $\Delta_{o,n}$ are anti-isomorphic.

(iv) $\Delta_{u,n} \kappa_A = \Delta_{o,n}$
In particular, the K-algebras $\Delta_{u,n}$ and $\Delta_{o,n}$ are isomorphic.

(v) The restriction of $\kappa_A \tau$ to $\Delta_{u,n}$ resp. to $\Delta_{o,n}$ is a $\mathcal{A}_1$-anti-automorphism of $\Delta_{u,n}$ resp. of $\Delta_{o,n}$ of order 2.
In particular, the K-algebras $\Delta_{o,n}$ and $\Delta_{u,n}$ are isomorphic to its inverse algebra.

Proof. ad (i): A straightforward calculation implies that A is self-inverse and symmetric. Thus, part (i) is proven.

ad(ii): For all $M \in K^{n \times n}$ we use part (i) to deduce

$$M\kappa_A \tau = (AMA)^t = A^t M^t A^t = AM^t A = M\tau \kappa_A.$$

Hence, part (ii) is proven.

ad(iii): This part is straightforward to calculate.

ad(iv): By using part (iii) and a dimension argument we have only to prove one inclusion. Let $M \in \Delta_{u,n}$. For all $i,j \in \underline{n}$ we calculate $(i;j)AMA = \sum_{k=1}^{n} \sum_{s=1}^{n} a_{is} m_{sk} a_{kj}$. In addition, $a_{ij} = 1_K$ resp. $a_{ij} = 0_K$ is valid if and only $i = n - j + 1$ resp. $i \neq n - j + 1$ is true. We deduce $(i;j)AMA = m_{(n-i+1)(n-j+1)}$. If $i < j$ is valid, then $n - i + 1 > n - j + 1$ is true. Because of $M \in \Delta_{u,n}$ we deduce $m_{(n-i+1)(n-j+1)} = 0_K$, and hence $AMA \in \Delta_{o,n}$ is valid.

ad(v): This part is deductable by the parts (ii), (iii) and (iv).◇

We will analyze the K-algebra $\Delta_{u,n}$. For every $r \in rad(\Delta_{u,n})$ we have to determine the K-algebra $D(n,K)^{1_{\Delta_{u,n}}+r}$. For this, we have to calculate the inverse matrix of $1_{\Delta_{u,n}} + r$. The inverse matrix can be determined by applying the next recursion method. In section 3.3.1 another approach for calculating the inverse matrix is presented based on the star composition and the nilpotency class of r.

Construction 1 *(radical complements of the algebra of lower triangular matrices)* Let K be a field, $n \in \mathbb{N}_{\geq 2}$, $M := \begin{pmatrix} 1_K & 0_K & \cdots & 0_K & 0_K \\ a_{2,1} & 1_K & \cdots & 0_K & 0_K \\ \cdot & \cdot & \cdots & \cdot & \cdot \\ a_{n-1,1} & a_{n-1,2} & \cdots & 1_K & 0_K \\ a_{n,1} & a_{n,2} & \cdots & a_{n,n-1} & 1_K \end{pmatrix} \in K^{n \times n}$,

$$A := \begin{pmatrix} 1_K & 0_K & \cdots & 0_K \\ a_{2,1} & 1_K & \cdots & 0_K \\ \cdot & \cdot & \cdots & \cdot \\ a_{n-1,1} & a_{n-1,2} & \cdots & 1_K \end{pmatrix} \in K^{(n-1)\times(n-1)},$$

$$B := \begin{pmatrix} 0_K & 0_K & \cdots & 0_K \end{pmatrix} \in K^{1\times(n-1)},$$

$$C := \begin{pmatrix} a_{n,1} & a_{n,2} & \cdots & a_{n,n-1} \end{pmatrix} \in K^{1\times(n-1)} \text{ and}$$

$D := (1_K) \in K^{1\times 1}$. In imprecise, but comfortable notation $M = \begin{pmatrix} A & B^t \\ C & D \end{pmatrix}$ is valid.

By using theorem 2.11 in [59] we deduce
$M^{-1} = \begin{pmatrix} X & Y \\ U & V \end{pmatrix}$ in which

$X = A^{-1} + A^{-1}B^t(D - CA^{-1}B^t)^{-1}CA^{-1}$,
$Y = -A^{-1}B^t(D - CA^{-1}B^t)^{-1}$,
$U = -(D - CA^{-1}B^t)^{-1}CA^{-1}$ and
$V = (D - CA^{-1}B^t)^{-1}$ are valid.

We conclude $X = A^{-1}$, $Y = 0_{K^{(n-1)\times 1}}$, $U = -CA^{-1}$ and $Y = D$. Thus, we have determined a recursion formula for determining the inverse matrix of M which is: $M^{-1} = \begin{pmatrix} A^{-1} & 0_{K^{(n-1)\times 1}} \\ -CA^{-1} & (1_K) \end{pmatrix}$.

If $r \in rad(\Delta_{u,n})$ and the inverse of $1_{\Delta_{u,n}} + r$ is calculated by the recursion method, then we have to calculate $D(n,K)^{1_{\Delta_{u,n}}+r}$. Let for all $i \in \underline{n}$ the element e_i be the $n \times n$-matrix over K such that the $(i;i)$-entry is 1_K and all other entries are zero. The set $\{e_i \mid i \in \underline{n}\}$ is a K-basis of $D(n,K)$. Thus, $\{e_i{}^{1_{\Delta_{u,n}}+r} \mid i \in \underline{n}\}$ is a K-basis of $D(n,K)^{1_{\Delta_{u,n}}+r}$. We have only to calculate $e_i{}^{1_{\Delta_{u,n}}+r}$ for every $i \in \underline{n}$. Let $i \in \underline{n}$. if we define $C :=$

$$\begin{pmatrix} (1_{\Delta_{u,n}}+r)_{i,1}(1_{\Delta_{u,n}}+r)^{-1}_{i,i} & \cdots & (1_{\Delta_{u,n}}+r)_{i,i}(1_{\Delta_{u,n}}+r)^{-1}_{i,i} \\ \cdots & \cdots & \cdots \\ (1_{\Delta_{u,n}}+r)_{i,1}(1_{\Delta_{u,n}}+r)^{-1}_{n,i} & \cdots & (1_{\Delta_{u,n}}+r)_{i,i}(1_{\Delta_{u,n}}+r)^{-1}_{n,i} \end{pmatrix} \in K^{(n-i+1)\times i},$$

$A := 0_{K^{(i-1)\times i}}$, $B := 0_{K^{(i-1)\times(n-i)}}$ and $D := 0_{K^{(n-i+1)\times(n-i)}}$, then a straightforward calculation implies:

$$e_i{}^{1_{\Delta_{u,n}}+r} = \begin{pmatrix} A & B \\ C & D \end{pmatrix}.$$

Finally, we remark that the algebra within example 1 is isomorphic to $\Delta_{u,2}$: if we define $I := \begin{pmatrix} 1_K & 0_K \\ 0_K & 1_K \end{pmatrix}$, $E := \begin{pmatrix} 0_K & 0_K \\ 0_K & 1_K \end{pmatrix}$ and $R := \begin{pmatrix} 0_K & 0_K \\ 1_K & 0_K \end{pmatrix}$, then $R^2 = 0_{K^{2\times 2}}, E^2 = E, ER = R$ and $RE = 0_{K^{2\times 2}}$ are valid. This implies that both algebras are isomorphic (see the multiplication matrix within example 1).

1.3.3 Closed shapes and matrix algebras

As stated earlier in this work we present some results of the work of Ian Stewart and Martin Golubitsky within [38] which are related to the theorem of Wedderburn-Malcev. These results are used within their analysis of dynamical networks.

Definitions 3 Let K be a field, $n \in \mathbb{N}$ and $M \in K^{n \times n}$. A subset S of $\underline{n} \times \underline{n}$ is called a shape or form of length n if and only if for all $i \in \underline{n}$ the condition $(i;i) \in S$ is valid. In other words, S is a reflexive relation on $\underline{n}$. A shape S is called closed if and only if for all $i, j, k \in \underline{n}$ satisfying $(i;j), (j;k) \in S$ the condition $(i;k) \in S$ is true. In other words, S is a transitive relation. M is of shape S if and only if for all $(i;j) \in \underline{n} \times \underline{n}$ the condition $M_{i,j} = 0$ is equivalent to $(i;j) \notin S$. By $K^{n \times n}(S)$ we denote the set of all matrices of $K^{n \times n}$ of shape S. A shape S is visualized by Stewart and Golubitsky by a so-called symbolic matrix for which the $(i;j)$-entry is $\star$ resp. 0 if and only if $(i;j)$ is contained resp. not contained in S.⋄

Example 2 Let K be a field and $n := 3$. We focus on the symbolic matrices
$\begin{pmatrix} \star & 0 & 0 \\ \star & \star & 0 \\ 0 & \star & \star \end{pmatrix}$ and $\begin{pmatrix} \star & 0 & 0 \\ 0 & \star & 0 \\ \star & 0 & \star \end{pmatrix}$. The first resp. second symbolic matrix is related to a closed resp. to a non-closed shape. ⋄

Closed shapes lead to unital subalgebras of $K^{n \times n}$ which is proven by Stewart und Golubitsky in lemma 5.6 of their work [38] by using elementary matrix calculations (which the reader may also verify as an exercise):

Theorem 9 *(Stewart, Golubitsky, 2015)* Let K be a field, $n \in \mathbb{N}$ and S a shape of length n. $K^{n \times n}(S)$ is a K-subspace containing the unit matrix. The set of elementary matrices $E_{i,j}$, $(i;j) \in S$ is a K-basis of $K^{n \times n}(S)$. Thus, this subspace is of dimension $\mid S \mid$. $K^{n \times n}(S)$ is an unital subalgebra of $K^{n \times n}$ if and only if S is closed.⋄

Within the analysis of dynamical networks closed shapes arise in the context of so-called cores. The reader may study the corresponding chapter 2 of the article [38] for this connection. Shapes can be used to create new shapes (see definition 12.1 and chapter 13 in [38]):

Definitions and remarks 1 Let $n \in \mathbb{N}$, K a field and S a shape of length n. We define $S^\tau := \{(j;i) \mid (i;j) \in S\}$. S^τ is a shape, too, and is called the dual shape of S. S is closed if and only if S^τ is closed. If S is closed, then $K^{n \times n}(S^\tau) = K^{n \times n}(S)^\tau$ is valid. Let $\pi \in S_n$. We define $S^\pi := \{(i\pi;j\pi) \mid (i;j) \in S\}$. S^π is a shape, too, and this shape is closed if and only if S is closed. The symmetric group of order n acts on the shapes of length n, and the subset of closed shapes is S_n-invariant. Let P_π the permutation matrix connected to π in $K^{n\times n}$. By using the basis transformation law we deduce that $K^{n\times n}(S^\pi) = P_\pi^{-1} \cdot K^{n\times n}(S) \cdot P_\pi$ is valid: the corresponding subalgebras $K^{n\times n}(S^\pi)$ and $K^{n\times n}(S)$ are conjugated.⋄

The following property of closed shapes is essential for the study of dynamical networks and let the theorem of Wedderburn-Malcev be applicable to the theory: based on a closed shape a equivalence relation is defined which is a partial order. This is a general method to obtain a partial order based on a reflexive and transitive relation. The equivalence classes linked to the partial order are called clusters by Stewart und Golubitsky. It is possible to permute the elements of $\underline{n}$ to obtain a so-called block-triangular form. The following theorem is valid (see theorem 8.8 in [38]):

Theorem 10 *(Stewart, Golubitsky, 2015) Let K be a field, $n \in \mathbb{N}$ and S a closed shape of length n. Elements $\pi \in S_n$, $p \in \mathbb{N}$ and $n_1, \cdots, n_p \in \mathbb{N}$ exist such that $K^{n\times n}(S)$ is – up to conjugation with P_π – block-triangular:*

$$\begin{pmatrix} K^{n_1\times n_1} & 0 & 0 & \cdots & 0 \\ \star & K^{n_2\times n_2} & 0 & \cdots & 0 \\ \vdots & \vdots & \vdots & \ddots & \vdots \\ \star & \star & \star & \cdots & K^{n_p\times n_p} \end{pmatrix}.$$
⋄

Those algebras possess a nilradical and a radical complement which can be determined from this block-triangular form: the nilradical is exactly the set of all matrices of block-triangular shape below the block-diagonal. A radical complement is the block-diagonal, and this complement is isomorphic to $\bigoplus_{i=1}^{p} K^{n_i\times n_i}$. The direct product of full matrix algebras over a base field is separable by using part (v) of corollary 1. Hence, the theorem of Wedderburn-Malcev can be applied to those algebras.

1.4 Invariant radical complements and Taft's theorem

We begin this section by presenting some examples for the context of Taft's theorem.

Examples 3 *(examples of group actions on radical complements)* (1) Let K be a field of characteristic $p > 0$ and G a finite group possessing a normal p-Sylow subgroup P. By using the theorem of Schur-Zassenhaus a complement H of P in G exists. Let us focus on the projection from G onto G/P and extend it K-linear to an epimorphism from KG onto $K(G/P)$. The kernel of this epimorphism is well-known: it is $KG \cdot Aug(KP) = Aug(KN) \cdot KG$ in which $Aug(KP)$ is the so-called augmentation ideal of KP. We use a theorem of Wallace to deduce that this ideal is nilpotent. Because of $KG \cdot Aug(KP) = Aug(KP) \cdot KG$ the kernel is nilpotent, too. P is a Hall subgroup, and thus $K(G/P)$ is separable based on theorem 2. We have proven $rad(KG) = Aug(KP) \cdot KG$. $K(G/P)$ is isomorphic to KH, and thus KH is a radical complement in KG. The function $g \mapsto g^{-1}$ extends linear to an involution ι of KG. The radical complement KH is invariant under ι. Because of $\iota\iota = id$ the set $X := \{\iota, id\}$ is a group of order 2 acting on KG. KH is invariant under the group action of X on KG. Is every radical complement invariant under the action of G? Which are the G-invariant radical complements?

If we consider an arbitrary finite group G, then Karpilovsky has proven in [22] that $KG/rad(KG)$ is always separable. Thus, by applying the Wedderburn-Malcev theorem a radical complement C exists. In view of the previous context a natural question is whether C or another or every complement is invariant under the action of G. How can we describe all G-invariant radical complements? Does always a radical complement exist which is isomorphic to a group algebra or which is a group algebra based on a subgroup of G?

(2) Let A be an associative unitary finite-dimensional commutative K-algebra over a field K. If $A/rad(A)$ is separable, then A possesses exactly one radical complement because of the Wedderburn-Malcev theorem (conjugation acts as identity for commutative algebras). Hence, this unique radical complement is invariant under all auto- (and anti-)automorphism of A.

(3) Let K be a field, $A := \begin{pmatrix} 1_K & 0_K & 1_K \\ 0_K & 1_K & 0_K \\ 1_K & 0_K & 1_K \end{pmatrix}$ and $B := \begin{pmatrix} 1_K & 0_K & 0_K \\ 0_K & 0_K & 0_K \\ 0_K & 0_K & 0_K \end{pmatrix}$.
Straightforward to calculate are $char_{A,K} = min_{A,K} = (1-t)(t-2)t$ and $char_{B,K} = min_{A,K}$. A, B are invariant under the anti-automorphism ^{t} : $X \mapsto X^t$ (the transpose matrix). Does a radical complement exists for $K[A, B]$? Does a radical complement exists for $K[A, B]$ which is invariant under ^{t}? Can we describe all radical complements of $K[A, B]$? Can we describe all radical complements of $K[A, B]$ which are invariant under ^{t}? In this example $\{id,^t\}$ is a group of order two acting on $K[A, B]$.

(4) Example (3) can be generalized to the subalgebra of $K^{3\times 3}$ (K a field) presented by the matrices of the form $\begin{pmatrix} a & 0_K & d \\ 0_K & b & 0_K \\ e & 0_K & c \end{pmatrix}$. This subalgebra is invariant under $\{id,^t\}$.

(5) A further generalization of example (4) are subalgebras of $K^{n\times n}$ ($n \in \mathbb{N}$,

K a field) which are invariant under $\{id,^t\}$. These are e.g. subalgebras of the form $D + N + N^t$ such that N is a subalgebra of strict lower (or upper) triangular matrices. Examples for N are the powers of the subalgebra of strict lower (or upper) triangular matrices. Also the subalgebras $D + N + N^t$ are invariant under $\{id,^t\}$.

(6) Let A be an associative unitary finite-dimensional K-algebra over a field K of characteristic $p > 0$ such that $A/rad(A)$ is separable. Take any element r of the nilradical $rad(A)$. Let $k \in \mathbb{N}$ minimal such that $cl(r) \leq p^k$. It is straightforward to calculate that $(\kappa_{1+r})^{p^k} = id$ is valid. Thus, $G := \langle \kappa_{1+r} \rangle_G$ is a p-group acting on the radical complements of A. Does a G-invariant radical complement exists? How can we describe all G-invariant radical complements of A?

(7) Let A be an associative unitary finite-dimensional K-algebra over a field K and $r \in rad(A)$ such that $r^2 = 0$ is valid. For all $n \in \mathbb{N}$ the rule $(\kappa_{1+r})^n = \kappa_{1+nr}$ is valid. The group $G := \langle \kappa_{1+r} \rangle_G$ is acting on the radical complements of A. Does a G-invariant radical complement exists? How can we describe all G-invariant radical complements of A?

(8) This example is based on an example presented by Taft in [46]. Let K be a field, $n \in \mathbb{N}$, x a primitive nth root of unity and $D = (d_{i,j})$ a diagonal matrix of $K^{n \times n}$ such that $d_{1,1} = x$ and $d_{i,i} = 1$ for all $2 \leq i \leq n$. The group G generated by κ_D is a cyclic group of order n. G is acting on the subalgebra of lower (or upper) triangular matrices. Of course, the subalgebra of diagonal matrices is a G-invariant radical complement. How can we describe all G-invariant radical complements? G is acting also on the subalgebras presented within parts (3), (4) and (5). Does a G-invariant radical complement exists and how can we describe all G-invariant radical complements?⋄

For presenting Taft's theorem we need some definitions and remarks.

Definition and remark 4 Let K be a field, A a finite-dimensional unitary associative K-algebra and G a finite group such that all elements of G are acting as auto- resp. anti-automorphism on G by $a.g := ag$ resp. $a.g = -ag$ for all $a \in A$ and $g \in G$. Let $a \in A$ and $x \in E(A)$. The element a is called G-skew (or G-symmetric) if $a.g = a$ resp. $a.g = -a$ for all auto- resp. anti-automorphism $g \in G$. The unit x is called G-orthogonal if $x.g = x$ resp. $x.g = -x^{-1}$ for all auto- resp. anti-automorphism $g \in G$. The collection of all G-orthogonal elements form a subgroup of $E(A)$ which acts on the G-invariant subalgebras and radical complements. As a consequence the set of all G-orthogonal elements of $1 + rad(A)$ form a subgroup of $1 + rad(A)$ which acts on the G-invariant subalgebras and radical complements.

Some of the questions within the previous example 3 can be answered by the following theorem of Taft:

Theorem 11 *(theorem of Taft for invariant radical complements)* Let K be a field, A a finite-dimensional unitary associative K-algebra possessing a separable factor algebra by its nilradical and G a finite group such that the characteristic of K is not dividing the order of G. All elements of G are acting as auto- resp. anti-automorphism on G by $a.g := ag$ resp. $a.g = -ag$ for all $a \in A$ and $g \in G$. The following statements are valid:

(i) *Existence part:* A possesses a G-invariant radical complement.

(ii) *Uniqueness part:* If $char(K) \neq 2$, then the group of G-orthogonal elements within $1 + rad(A)$ acts transitive on the set of all G-invariant radical complements of A by conjugation.

Proof. We present a proof within the appendix (see chapter C) based on Taft's articles [41], [42], [43], [44], [45] and [46].◇

The reader may analyze the examples 3 in view of Taft's theorem as an exercise (see exercise 23). Nevertheless, we present one result related to these examples for group algebras:

Corollary 1 *(group algebras and invariant radical complements by inverting)* Let G be a finite group and K a field. KG possesses a radical complement and all radical complements are conjugated under the action of $1 + rad(KG)$ by conjugation. In addition, KG possesses an invariant radical complement under the action of the linearization $\iota : g \mapsto g^{-1}$. Let $G := \{id, \iota\}$. G is a group of order 2. If $char(K) \neq 2$, then all G-invariant radical complements are G-orthogonal conjugated. Every separable subalgebra can be G-orthogonal conjugated into a G-invariant radical complement.

Proof. This corollary is derived directly by using the examples 3, theorem 11 and the next corollary 2.◇

As proven within the context of the Wedderburn-Malcev theorem the following result is valid for G-invariant separable subalgebras:

Corollary 2 *(maximality of invariant radical complements)* Let K be a field, A a finite-dimensional unitary associative K-algebra possessing a separable factor algebra by its nilradical and G a finite group such that the characteristic of K is not dividing the order of G. All elements of G are acting as auto- resp. anti-automorphism on G by $a.g := ag$ resp. $a.g = -ag$ for all $a \in A$ and $g \in G$. If $char(K) \neq 2$, then every G-invariant separable subalgebra T of A can be conjugated G-orthogonally based on an elements within $1 + rad(A)$ into a G-invariant radical complement of A.

Proof. The proof is based on the argumentation within the proof of corollary 4. Let S be a G-invariant radical complement. We consider the subalgebra $B := rad(A) \oplus S$. It is proven that $T \cap B$ is another radical complement within B. S is G-invariant and $T \cap B$ also because T (by our assumption) and $rad(A)$ (because the nilradical is invariant under auto- and anti-automorphism) are G-invariant. Now we can apply the uniqueness part of theorem 11 to S and $T \cap B$.⋄

We end this subsection with some remarks to the case of a finite algebra.

Remark 6 This remark is based on a conversation with Frieder Ladisch within mathoverflow (see [64]). Within the assumption of the theorem of Taft 11 let K be a finite field of characteristic $p > 0$. Then the algebra A is finite and the group $1+rad(A)$ is a p-group (because $1+rad(A)$ has the same order as $rad(A)$ which is a K-space) acting transitive on the set of radical complements because of the Wedderburn-Malcev theorem. In this case K is perfect and therefor $A/rad(A)$ is separable. Because of the transitive action of $1 + rad(A)$ the set of radical complements is of p-power order, too. The situation described within Taft's theorem 11 is generalized (for the case G possesses only automorphism) by the following theorem of Glauberman (see [12]) within the context of group actions:

Theorem (Glauberman). Let the finite group G act on the finite group N by automorphisms, where $(|G|, |N|) = 1$. Let both G and N act on a set Ω, where the action of N is transitive, and the compability condition

$$(\omega n)g = (\omega g)n^g \quad \text{for all } \omega \in \Omega, \, n \in N, \, g \in G$$

holds. Then Ω has a fixed point under G. Moreover, $C_N(G)$ acts transitively on the G-fixed points.

This theorem was generalized to a finite group possessing also anti-automorphism by Taft within his work [47].⋄

1.5 Connections to the theorems of Schur-Zassenhaus and Levi-Malcev

The theorem of Wedderburn-Malcev within the theory of associative algebras has a pendant within group and Lie algebra theory.

The theorem of Schur-Zassenhaus indicates that every normal Hall subgroup (a normal subgroup possessing an order coprime to the order of its factor group within the whole group) of a finite group possesses a complement. If the normal Hall subgroup or its factor group is solvable, then all its complements are conjugated. Based on the odd-order theorem of Feit-Thompson the normal Hall subgroup or its factor group is solvable. Thus, we can say that the separability within associative algebras is replaced by coprime orders. If modern proofs are analyzed for both theorems based on cohomology theory, then further connections are visible: the existence part of the theorem of Wedderburn-Malcev is reduced to a zero nilradical and the one for the theorem of Schur-Zassenhaus to an Abelian normal Hall subgroup.

The theorem of Levi-Malcev is based on finite-dimensional Lie algebras over fields of characteristic 0: the solvable radical possesses a complement and all complements are conjugated by an automorphism of the form $exp(ad(x))$ such that x is contained in the nilpotent radical. Connections are given by using characteristic zero instead of the separability. A proof based on cohomology theory is reduced to the case of an Abelian Lie algebra.

K.W. Roggenkamp analyzed the cohomology concept in his work of the seminar series [36] and developed a common cohomology theory such that all three theorems can be deduced from it.

In what follows, we want to state some more connections between the theorem of Wedderburn-Malcev and the theorem of Schur-Zassenhaus. We ask whether it is possible to derive the theorems from each other. To connect group theory and associative algebra theory we use the group algebra and the group of units.

For applying the theorem of Schur-Zassenhaus we need finite groups. Thus, we want to deduce the theorem of Wedderburn-Malcev for associative algebras over finite fields. (Finite fields are perfect. Hence, the factor algebra by the nilradical is separable and we could also apply the theorem of Wedderburn-Malcev directly.) Let us focus on a finite-dimensional associative unitary K-algebra A over a field K of order p^r such that p is a prime number and $r \in \mathbb{N}$. Finite division rings are commutative based on a theorem of Jacobson. Hence, we derive by using the theorem of Wedderburn-Artin for the structure of associative algebras that $s \in \mathbb{N}$, $n_1, \cdots, n_s \in \mathbb{N}$ and field extensions $K_1, \cdots, K_s$ of K exist such that $A/rad(A)$ and $\bigoplus_{i=1}^{s} K_i^{n_i \times n_i}$ are isomorphic. $E(A)/(1 + rad(A)) = E(A/rad(A))$ is valid (see lemma 10 within the appendix). It is straightforward to prove that $\mid rad(A) \mid$ is a p-power and $E(A)/(1+rad(A))$ is a $p^{'}$-group. Within the group $E(A)$ the set $1+rad(A)$ is a normal Hall subgroup possessing a complement U by applying the theorem of Schur-Zassenhaus. We calculate $\langle E(A) \rangle_K = rad(A) \oplus \langle U \rangle_K$. This set is an unital subalgebra. If S, T are two algebra complements of $rad(A)$ in A, then they are unital and their group of units are complements

of $1 + rad(A)$ in $E(A)$. Hence, $E(S)$ and $E(T)$ are conjugated based on the theorem of Schur-Zassenhaus. Thus, their K-spans are conjugated, too. If the stated K-spans would be the algebra again, then we would have proven the theorem. In our case we need to restrict this statement to semisimple algebras and within this class to the algebra $L^{n \times n}$ such that $n \in \mathbb{N}$ and L is a finite field extension of the finite base field K. Elementary transformations of columns and rows are representable by using a multiplication with invertible matrices from left resp. right. Thus, we need to focus on the matrix
$\begin{pmatrix} b & 0 & 0 & \cdots & 0 \\ 0 & 0 & 0 & \cdots & 0 \\ \vdots & \vdots & \vdots & \ddots & \vdots \\ 0 & 0 & 0 & \cdots & 0 \end{pmatrix}$. Here, b is an element of L different from zero. This matrix is the product of $\begin{pmatrix} 1 & 0 & 0 & \cdots & 0 \\ 0 & 0 & 0 & \cdots & 0 \\ \vdots & \vdots & \vdots & \ddots & \vdots \\ 0 & 0 & 0 & \cdots & 0 \end{pmatrix}$ with $b \cdot I_{n \times n}$ ($I_{n \times n}$ is the unity matrix.). The latter matrix is invertible. If L possesses at least three elements, then we choose an element $a \in L$ different from 1 and 0. The matrix $\begin{pmatrix} 1 & 0 & 0 & \cdots & 0 \\ 0 & 0 & 0 & \cdots & 0 \\ \vdots & \vdots & \vdots & \ddots & \vdots \\ 0 & 0 & 0 & \cdots & 0 \end{pmatrix}$ is exactly the sum of $\begin{pmatrix} a & 0 & 0 & \cdots & 0 \\ 0 & a & 0 & \cdots & 0 \\ \vdots & \vdots & \ddots & \vdots & \vdots \\ 0 & 0 & 0 & \cdots & a \end{pmatrix}$ and $\begin{pmatrix} 1-a & 0 & 0 & \cdots & 0 \\ 0 & -a & 0 & \cdots & 0 \\ \vdots & \vdots & \ddots & \vdots & \\ 0 & 0 & 0 & \cdots & -a \end{pmatrix}$. Both summands are invertible. Thus, we have proven our statement for fields possessing at least three elements.

The opposite statement – deducing the theorem of Schur-Zassenhaus from the theorem of Wedderburn-Malcev – can not be proven here. But the author has the following idea for proving it: Let G be a finite group possessing a normal Hall subgroup N. As stated earlier, the proof can be reduced to the case that N is Abelian (and by a more deeper analysis) and a p-group. We choose a finite field of characteristic p, like $K = GF(p)$. Let us focus on the projection from G onto G/N and extend it K-linear to a epimorphism from KG onto $K(G/N)$. The kernel of this epimorphism is well-known: it is $KG \cdot Aug(KN) = Aug(KN) \cdot KG$ in which $Aug(KN)$ is the so-called augmentation ideal of KN. We use a theorem of Wallace to deduce that this ideal is nilpotent. Because of $KG \cdot Aug(KN) = Aug(KN) \cdot KG$ the kernel is nilpotent, too. N is a Hall subgroup, and thus $K(G/N)$ is separable based on theorem 2. We have proven $rad(KG) = Aug(KN) \cdot KG$. Now we apply the theorem of Wedderburn-Malcev to deduce that $rad(KG)$ possesses a

Separable algebras and the theorem of Wedderburn-Malcev 45

complement T in A. We need to prove that an element $j \in rad(KG)$ exists such that $E(T^{1+j}) \cap G$ is a complement of the normal Hall subgroup N of G. The proof is not known to the author. If we use the Schur-Zassenhaus theorem and apply a complement U of N in G to KG, then KU is a separable radical complement in KG. The intersection of KU with G is KU.⋄

1.6 Open-ended questions and exercises

Open-ended question 1 *(i) Find connections between the theorems of 'Wedderburn-Malcev', 'Schur-Zassenhaus' and 'Levi-Malcev'.*

(ii) Describe those finite-dimensional associative algebras A such that $Der(A) = Inder(A)$ is valid.

Excercise 1 *Let K be a field and A be an associative simple finite-dimensional K-algebra. Prove that $Z(A)$ is a field and A an associative simple finite-dimensional central $Z(A)$-algebra. If every derivation of A as K-algebra is zero on $Z(A)$, then A possesses only inner derivations as K-algebra. Is it possible to generalize this statement to semisimple algebras? Is it possible to generalize this statement to arbitrary (A,A)-bimodules M (instead of considering only the case $M = A$)?*

Excercise 2 *Prove remark 2 in details.*

Excercise 3 *Solve exercise 2 on page 209 in [35]. The exercise presents an alternative proof that factors sets of bimodules for separable algebras are splitting (based on the property that all derivations are inner derivations).*

Excercise 4 *Generalize the construction 1 to an arbitrary lower (and upper) triangular matrix! Calculate examples in three and four dimensions. (Hint: The algebra decompose into $s\delta_{u,n} := N$ and $D(n,K) := D$. Hence, its group of units is the semidirect product of $1 + N$ and $E(D)$. A lower triangular matrix is invertible if and only if all entries on its diagonal are non-zero. Write $a = d + n = (1 + nd^{-1})d$. The inverse of d is straightforward to calculate. Now $(d+n)^{-1} = d^{-1}(1+nd^{-1})^{-1}$. Apply the construction already proven for $1 + N$ to the second factor.)*

Excercise 5 *Prove that A is separable if and only if A^{op} is separable by using theorem 1.*

Excercise 6 *Prove that A is separable if and only if $A \otimes A^{op}$ is separable by using theorem 1, exercise 5 and exercise 287.*

Excercise 7 *Prove that A is separable if and only if for every field extension $(K; L)$ the L-algebra $A_L := A \otimes_K L$ (basic field or scalar extension) is separable. (Hint: use theorem 1 and focus on separating idempotents under scalar extensions)*

Excercise 8 Prove that A is separable if and only if A is projective as (A, A)-bimodule. Find a connection between (A, A)-bimodules and $A \otimes_K A^{op}$-modules.

Excercise 9 Prove that A is separable if and only if a finite and separable field extension $(K; L)$ exists such that $A \otimes_K L$ is isomorphic to a direct sum of full matrix algebras over E. E is called a separable splitting field for A. If the exercise is to complex, then do a research in the literature to solve it.

Excercise 10 Do a research in the literature and find a connection between separable algebras and Frobenius algebras (Hint: theorem of Higman, [6]).

Excercise 11 Do a research in the literature and find a connection between separable algebras and flat modules over $A \otimes A^{op}$.

Excercise 12 Do a research in the literature and find a connection between separable algebras and the fact that the trace form $(a; b) \mapsto tr(ab)$ is non-degenerate.

Excercise 13 True or false: A is separable if and only if the multiplication map $a \otimes b \mapsto ab$ has a right inverse.

Excercise 14 True or false: A is separable if and only if the multiplication map $a \otimes b \mapsto ab$ has a left inverse.

Excercise 15 Do a research in the literature and find the definition of strongly separable algebras. What is the connection to separable algebras?

Excercise 16 Let G be a finite group consisting of auto- and anti-automorphism acting on an associative K-algebra A. Prove that the collection of automorphism is a subgroup of G. If G possesses an anti-automorphism, then the order of G is even. Furthermore, the collections of auto- resp. anti-automorphism of G have the same order.

Excercise 17 Let r be a G-skew element within the nilradical of an associative algebra A. True or false:

(i) $1 + r$ is G-orthogonal.

(ii) $1 - r$ is G-orthogonal.

(iii) $(1 + r)^{-1}$ is G-orthogonal.

(iv) $(1 - r)^{-1}$ is G-orthogonal.

(v) $(1 + r)^{-1}(1 - r)$ is G-orthogonal.

Separable algebras and the theorem of Wedderburn-Malcev 47

(vi) $(1-r)^{-1}(1+r)$ *is G-orthogonal.*

Excercise 18 *Let r be an element within the nilradical of an associative algebra A. Prove that $1 + (1-r)(1+r)^{-1} = 2(1+r)^{-1}$ is an unit and that $r = (1-(1-r)(1+r)^{-1})(1+(1-r)(1+r)^{-1})^{-1}$ is valid. In addition, prove that r is G-skew if and only if $(1-r)(1+r)^{-1}$ is G-orthogonal. Is the set of G-orthogonal elements $(1-r)(1+r)^{-1}$ a subgroup or normal subgroup of $E(A)$?*

Excercise 19 *Prove that the set of G-skew elements of an associative A form a Lie subalgebra in the associated Lie algebra A° of A with respect to the multiplication $a \circ b := ab - ba$ for all $a, b \in A$.*

Excercise 20 *Construct an example of a group of order 6 acting on a set M of order 5 possessing two orbits of length 2 and 3.*

Excercise 21 *Let $1 \neq U, V \neq S_3$ be subgroups of S_3. Consider the action of U on the right or left cosets of G with respect to V. Prove that a fixed point under this action exists.*

Excercise 22 *Let G be a finite p-group acting on a finite set M such that p does not divide the order of M. Prove that M possesses a fixed point under the action of G. Is this result true if $(|G|, |M|) = 1$ is valid?*

Excercise 23 *Analyze the examples 3 in view of Taft's theorem.*

Excercise 24 *Prove the statements within definitions and remarks 4.*

Excercise 25 *Let K be a field, $n \in \mathbb{N}$ and T a subalgebra of $\Delta_{o,n}$ containing the radical of $\Delta_{o,n}$. True or false: T possesses only inner derivations.*

Excercise 26 *Let K be a field, $n \in \mathbb{N}$ and T an unital subalgebra of $\Delta_{o,n}$ containing the radical of $\Delta_{o,n}$. True or false: T possesses only inner derivations.*

Excercise 27 *Let K be a field, $n \in \mathbb{N}$ and T a subalgebra of $\Delta_{o,n}$ containing $D(n, K)$. True or false: T possesses only inner derivations.*

Excercise 28 *Let K be a field and $n \in \mathbb{N}$. Prove that $\Delta_{u,n}$ possesses only inner derivations.*

Excercise 29 *Prove all statements within proposition 1.*

Excercise 30 *Let K be a field of characteristic zero, D_n ($n \in \mathbb{N}$) be a Solomon algebra over K and T a radical complement in A. If d is a derivation of such that $d_{|T} \neq 0$, then T is not d-invariant. Generalize this exercise to associative algebras possessing a commutative separable radical factor algebra and only inner derivations.*

Excercise 31 Let A be an associative unitary K-algebra and M an (A, A)-bimodule. $Der(A, M)$ is a Lie algebra by using the Lie multiplication $d \circ e := de - ed$ (where ed resp. de is the composition of functions). In addition, $Inder(A, M)$ is a Lie ideal of $Der(A, M)$. Determine $Inder(A, M)$ and $Inder(A)$ in more details related to A.

Excercise 32 Let K be a field and A an associative three-dimensional unitary K-algebra possessing the basis $\{1, e_2, e_3\}$ such that all three elements are idempotent and $e_2 e_3 = e_2$ and $e_3 e_2 = e_3$ are valid. Determine all derivations of A in A. This algebra is the Solomon-Tits algebra in dimension 3 symbolized by $K\Pi_3$. On what terms is $Der(A) = Inder(A)$ valid?

Excercise 33 Let A be an associative unitary K-algebra, M an (A, A)-bimodule, $d \in Der(A, M)$ and e an idempotent of A. Prove the following statements:

(i) If A is unitary, then $d(1) = 0$.

(ii) $d(e) = ed(e) + d(e)e$

(iii) If e is central, then $d(e) = 2ed(e)$ is valid.

(iv) If e is central, then $eA = Ae$ is d-invariant and unitary possessing the unit e.

(v) If e is central, then $d(e) = 0$.

Excercise 34 This exercise is taken from [8], chapter 6 and is adjusted. Let A be an associative algebra of finite-dimension and $d \in End_K(A)$. The function $\hat{d} : A \longrightarrow A^{2\times 2}, a \mapsto \begin{pmatrix} a & d(a) \\ 0 & a \end{pmatrix}$ is an algebra homomorphism if and only if d is a derivation of A. Now apply this result to the case of a central division algebra A and prove – by using the theorem of Skolem-Noether – that only inner derivations exists. Is it possible to derive this result by another argumentation?

Excercise 35 Let K be a field and A the algebra of strict upper triangular matrices in $K^{3\times 3}$. For the following functions d investigate whether they are derivations and inner derivations of A into A:

(i) $d : \begin{pmatrix} 0_K & a & b \\ 0_K & 0_K & c \\ 0_K & 0_K & 0_K \end{pmatrix} \mapsto \begin{pmatrix} 0_K & 0 & b \\ 0_K & 0_K & 0_K \\ 0_K & 0_K & 0_K \end{pmatrix}$

(ii) $d : \begin{pmatrix} 0_K & a & b \\ 0_K & 0_K & c \\ 0_K & 0_K & 0_K \end{pmatrix} \mapsto \begin{pmatrix} 0_K & 0_K & 0_K \\ 0_K & 0_K & c \\ 0_K & 0_K & 0_K \end{pmatrix}$

Separable algebras and the theorem of Wedderburn-Malcev   49

(iii) $d : \begin{pmatrix} 0_K & a & b \\ 0_K & 0_K & c \\ 0_K & 0_K & 0_K \end{pmatrix} \mapsto \begin{pmatrix} 0_K & a & 0_K \\ 0_K & 0_K & 0_K \\ 0_K & 0_K & 0_K \end{pmatrix}.$

In addition, decide whether the mentioned functions d are algebra-homomorphism or an element n exists such that for all $a, b \in A$ the rule $d(xy)^n = d(x)^n d(y)^n$ is valid. For which of the functions d this rule is valid for all $n \in \mathbb{N}$?

Excercise 36 Let A be an associative one-dimensional K-algebra. Determine all derivations of A into A. Analyze the cases of A being unitary and non-unitary.

Excercise 37 Let A be an associative two-dimensional K-algebra. Determine all derivations of A into A. Analyze the cases of A being unitary and non-unitary.

Excercise 38 Let A be an associative K-algebra. On what terms is $id_A \in Der(A)$ valid?

Excercise 39 For the cyclic groups G of order 2 and 3 describe all derivations of KG into KG based on a field K of arbitrary characteristics. Conjecture a result for cyclic group of prime order p and prove it.

Excercise 40 Let A be an associative unitary K-algebra, M an (A, A)-bimodule, $d \in Der(A, M)$, $a, x, t \in A$ and $m \in M$. Analyze the following statements for general M and for $M = A$:

(i) $d(1) = 0$

(ii) If x is invertible, then $d(x^{-1}) = -x^{-1}d(x)x^{-1}$.

(iii) Let x be invertible. $d(x)$ is invertible if and only if $d(x^{-1})$ is invertible. If $d(x)$ is invertible, then $d(x)^{-1} = -xd(x^{-1})x$.

(iv) Let x be invertible. Prove that $d(a^x) = (a.d(x).x^{-1} + d(a) - d(x).x.a)^x$ is valid.

Apply all valid formulas to $ad(m)$ resp. to $ad(t)$. Describe the kernel of $ad(m)$ resp. $ad(t)$ for a general M resp. for $M = A$.

Excercise 41 Let A be an associative K-algebra and $D \in Der(A)$. True or false: If D is bijective, then D^{-1} is a derivation of A. In addition, prove that D is not bijective if A is unitary or possesses a central idempotent different from zero.

Excercise 42 Let A be an associative K-algebra, M an (A, A)-bimodule, $D \in Der(A, M)$, $m \in M$, $n \in \mathbb{N}$ and $a, b \in A$. Determine $(ab)(D^n)$, $(ab)(ad(m)^n)$, $(a^n)D$ and $(a^n)ad(m)$. Analyze the determination for A being commutative and for $M = A$.

Excercise 43 Let A be an associative K-algebra, M an (A,A)-bimodule, $m \in M$, $D \in Der(A,M)$ and $char(K) = p > 0$. Prove $D^p \in Der(A,M)$. Describe $ad(m)^p$ in general and within the case $M = A$. Is every square of a derivation a derivation?

Excercise 44 True or false: The kernel of a derivation into a bimodule as an unital subalgebra. Determine the kernel of a inner derivation into a bimodule and into the algebra itself.

Excercise 45 Let A be an associative unitary K-algebra and $d \in Der(A)$. If e is an idempotent of A, then $d(e)^{2n+1} = d(e)$ is valid for all $n \in \mathbb{N}$. If r is a nilpotent element of class s, then $d(r)^{2s-1} = 0$. The set of nilpotent elements is invariant under d. If A is a basic algebra, then $rad(A)$ is d-invariant. Is $rad(A)$ d-invariant for arbitrary A?

Excercise 46 Prove lemma 3 within the article [24]. In what way is it possible to generalize this statement?

Excercise 47 Read the article [21] and analyze the importance of the theorem of Wedderburn-Malcev within the asymptotic of polynomial identities.

Excercise 48 Prove example 2 in details and specify the shapes explicitly.

Excercise 49 Let $n \in \mathbb{N}$ and K be a field. Is every unital subalgebra of $K^{n \times n}$ of the form $K^{n \times n}(S)$ based on a shape S of length n?

Excercise 50 Prove theorem 9 in details.

Excercise 51 Let $n \in \mathbb{N}$ and K be a field. Determine all shapes and all closed shapes of length n. Count these shapes!

Excercise 52 Solve exercise 51 for lower triangular shapes.

Excercise 53 Solve exercise 51 for upper triangular shapes.

Excercise 54 For the case $n = 3$ analyze the action of the symmetric group on the (closed) shapes. Determine the orbits and the number of the orbits in details.

Excercise 55 Prove definition and remark 1 in details.

Excercise 56 Analyze the section 1.3.2 about triangular matrices and transfer the results to unital subalgebras of $K^{n \times n}$ which are based on closed and lower or upper triangular shapes. A natural nilradical complement is the set of diagonal matrices, the nilradical is the set of corresponding strict upper or lower triangular matrices. All nilradical complements can be determined by using the mentioned algorithm. Apply this algorithm on the case $n = 7$ for one meaningful shape.

Separable algebras and the theorem of Wedderburn-Malcev 51

Excercise 57 Let K be a field, S a closed shape and $n \in \mathbb{N}$. Prove the statement $E(K^{n \times n}(S)) = K^{n \times n}(S) \cap E(K^{n \times n})$. In other words: If a matrix of closed shape S is invertible, then its inverse is of shape S. (Tip: direct proof or prove a general statement for finite dimensional unital subalgebras)

Excercise 58 Which closed shapes are total, injective, surjective, bijective, functions, anti symmetric?

Excercise 59 Apply to a closed shape of length 7 (which is not block-triangulated) a permutation such that the shape is block-triangulated afterwards. If needed, then use the statements of chapters 7 and 8 in [38].

Excercise 60 Is the inverse algebra of an associative separable algebra also separable?

Excercise 61 Let K be a field, $n \in \mathbb{N}$, A, B finite-dimensional associative unitary K-algebras, e a central idempotent of A, M a finite monoid and G a finite group. For the following algebras A decide whether the algebra and its inverse algebra are isomorphic:

(i) $\Delta_{u,n}$

(ii) $\Delta_{o,n}$

(iii) KM, for M being commutative and idempotent

(iv) KG, for G being Abelian

(v) KG, for $G = Q_8$ and $K := \mathbb{Q}$

(vi) KG, for $G = D_8$ and $K := \mathbb{R}$

(vii) KG, for $G = SD_8$ and $K := \mathbb{C}$

(viii) $K^{n \times n}$

(ix) $\mathbb{H}$

(x) eAe, for A being isomorphic to A^{op} (see exercise 173)

(xi) the zero extension of A, for A being separable and isomorphic to A^{op} (see exercise 174).

If the condition $A \cong A^{op}$ is not true within these examples, then analyze on which terms the condition it true.

Excercise 62 Within exercise 61 analyze whether all algebras are separable.

Excercise 63 Within exercise 61 determine the nilradical for all algebras.

Excercise 64 *Within exercise 61 analyze for all algebras whether the factor algebra by the nilradical is separable.*

Excercise 65 *Let K be a field, A a separable associative K-algebra and e an idempotent of A. Are eA and Ae separable? (Tip: matrix algebras)*

Excercise 66 *Let A be an associative K-algebra and e an idempotent of A. Is eAe exactly the intersection of eA and Ae? Is eAe an ideal of A? Is eAe an ideal of $eA + Ae$?*

Excercise 67 *Solve exercise 65 again for e being central.*

Excercise 68 *Prove corollary 1 in details.*

Excercise 69 *Prove remark 1 in details.*

Excercise 70 *Prove remark 2 in details.*

Excercise 71 *Do a research in the literature for remark 3.*

Excercise 72 *Let A be an associative K-algebra and $n \in \mathbb{N}$. Prove that $(A^{n \times n})^-$ and $(A^-)^{n \times n}$ are isomorphic.*

Excercise 73 *Let A be an associative K-algebra and $n, m \in \mathbb{N}$. To which algebra is $A^{n \times n} \otimes A^{m \times m}$ isomorphic?*

Excercise 74 *Let A, B associative K-algebras and $n, m \in \mathbb{N}$. To which algebra is $A^{n \times n} \otimes B^{m \times m}$ isomorphic?*

Excercise 75 *Let K be a field and G, H finite groups. If KG and KH are semisimple, then $K(G \times H)$ is semisimple, too. Is this statement true for separable group algebras?*

Excercise 76 *What is the consequence within the proof of remark 4, if the unit element is replaced by a general idempotent element? Analyze whether all idempotent elements are contained in one specific radical complement. Further analyze whether each idempotent element is contained in at least one radical complement.*

Excercise 77 *True or false: Powers of a field are subfields. If this statement is not true, then analyze on what terms it is true.*

Excercise 78 *Apply the construction 1 to the algebra $\Delta_{u,4}$ in the following way: at first determine a basis of the nilradical and determine the inverse matrix for each basis matrix shifted by the unity matrix. By using these inverse matrices calculate a basis for the conjugates of the radical complements of $D(4, K)$. How many radical complements arise in this way?*

Excercise 79 Transfer exercise 78 to the algebra $\Delta_{o,4}$!

Excercise 80 Let A be an associative finite-dimensional unitary K-algebra possessing a separable factor algebra by its nilradical and $n \in \mathbb{N}$. Within the matrix algebra $A^{n \times n}$ focus on the first column-space (which are those matrices over A possessing only values not equal to zero within the first column). Is this set a subalgebra? If this is true, then determine its nilradical. Is it possible to describe the nilradical by using the one of $A^{n \times n}$? Determine one radical complement of the first column-space. Is it possible to describe one radical complement by using one of $A^{n \times n}$? If this exercise is too complex, then solve it for special types of algebras A, like $A = K$ or A a field or skew field or A separable.

Excercise 81 Transfer exercise 80 to the algebra linked to the first column!

Excercise 82 Define an unitary associative algebra, an unitary and unital subalgebra and analyze the connections and differences between these topics (also by using suitable examples).

Excercise 83 Under the conditions of theorem 1 determine or characterize those kind of algebras for which each scalar extension is simple (and not only semisimple).

Excercise 84 Try to find a definition of a separable algebra module based on scalar extension and prove that the algebra modules are exactly the semisimple modules possessing a separable endomorphism algebra.

Excercise 85 Transfer exercise 83 to algebra modules.

Excercise 86 Analyze the connections between the topics 'nilpotent element', 'idempotent element' and 'unit'.

Excercise 87 Let us focus on the nilradical of the algebra $\Delta_{u,3}$. What is its class of nilpotency? Prove that the class of nilpotency for each element of the nilradical is not greater than the class of nilpotency of the nilradical. Does an element of the nilradical exist such that the class of nilpotency for this element is exactly the class of nilpotency of the nilradical?

Excercise 88 Solve exercise 87 for $\Delta_{u,4}$ and afterwards for $\Delta_{u,n}$ based on an arbitrary element $n \in \mathbb{N}$.

Excercise 89 Determine the following sets and their orders:

(i) $\underline{5} \setminus \underline{3}$

(ii) $\underline{17} \cap \underline{13}$

(iii) $P(\underline{3})$

(iv) $\underline{42} \cup \underline{111}$

(v) $\underline{2} \times \underline{3}$.

Generalize this exercise to arbitrary $n, m \in \mathbb{N}$!

Excercise 90 *True or false: Every semisimple algebra is separable.*

Excercise 91 *True or false: Every separable algebra is semisimple.*

Excercise 92 *The conjugation by an unit within an associative unitary K-algebra is K-linear.*

Excercise 93 *True or false: Every algebra is isomorphic to its inverse algebra.*

Excercise 94 *Why is the dimension of the real numbers as vector space over the rational numbers infinite?*

Excercise 95 *Let $A := \begin{pmatrix} 0_K & 0_K & 1_K \\ 0_K & 1_K & 0_K \\ 1_K & 0_K & 0_K \end{pmatrix}$ and $M := \begin{pmatrix} 1_K & 2_K & 3_K \\ 4_K & 5_K & 6_K \\ 7_K & 8_K & 9_K \end{pmatrix}$.
Show that A is invertible. Represent the conjugation with A by a matrix based on the standard basis of $K^{3\times 3}$ and calculate M^A. Is M invertible?*

Excercise 96 *Let A and M as in exercise 95. Calculate M^t, $(M^t)^{-1}$ and $((M^t)^{-1})^A$.*

Excercise 97 *True or false: $\Delta_{u,3}$ and $\Delta_{o,3}$ are anti-isomorphic. If the answer is yes, then represent this anti-isomorphism by a matrix based on two specified basis.*

Excercise 98 *True or false: $\Delta_{u,4}$ and $\Delta_{o,4}$ are isomorphic. If the answer is yes, then represent this isomorphism by a matrix based on two specified basis.*

Excercise 99 *The function taking the transpose matrix within $\mathbb{C}^{4\times 4}$ is $\mathbb{C}$-linear. Represent this function by a matrix based on a specified basis.*

Excercise 100 *Calculate the inverse matrix for the following matrices:*

(i) $\begin{pmatrix} 1_K & 0_K & 0_K & 0_K \\ 2_K & 1_K & 0_K & 0_K \\ 3_K & 4_K & 1_K & 0_K \\ 5_K & 6_K & 7_K & 1_K \end{pmatrix}$ *based on $K = GF(7)$*

Separable algebras and the theorem of Wedderburn-Malcev 55

(ii) $\begin{pmatrix} 1_K & 0_K & 0_K & 0_K \\ i_K & -1_K & 0_K & 0_K \\ -i_K & 1_K & 1_K & 0_K \\ 0_K & 0_K & (1+i)_K & 1_K \end{pmatrix}$ based on $K = \mathbb{Q}(i)$

(iii) $\begin{pmatrix} 1_K & 0_K & 0_K & 0_K \\ 2_K & 1_K & 0_K & 0_K \\ \sqrt{2}_K & -\sqrt{2}_K & 1_K & 0_K \\ 0.1_K & 6.5_K & \frac{1}{3}_K & 1_K \end{pmatrix}$ based on $K = \mathbb{R}$

(iv) $\begin{pmatrix} 1_K & 0_K & 0_K & 0_K \\ i_K & 1_K & 0_K & 0_K \\ \sqrt{2}i_K & 4_K & 1_K & 0_K \\ -0.1i_K & 6_K & \frac{16}{17}i_K & 1_K \end{pmatrix}$ based on $K = \mathbb{C}$.

Excercise 101 *Within exercise 100 calculate the conjugated subalgebra of the subalgebra of lower triangular matrices based on the given matrices.*

Excercise 102 *Let K be a field, $n \in \mathbb{N}$ and A an associative finite-dimensional K-algebra. True or false:*

(i) *If A is commutative and semisimple, then A is separable.*

(ii) *If $Z(A)$ is separable, then A is separable.*

(iii) *If $A^{n \times n}$ is semisimple, then A is semisimple.*

(iv) *If $A^{n \times n}$ is separable, then A is separable.*

(v) *If A possesses a radical complement, then $A/\mathrm{rad}(A)$ is separable.*

(vi) *If A possesses only conjugated radical complements based on the action of $1 + \mathrm{rad}(A)$, then $A/\mathrm{rad}(A)$ is separable.*

(vii) *Let T be a radical complement. The action of $E(A)$ on T by conjugation can be replaced by the action of $1 + \mathrm{rad}(A)$.*

Excercise 103 *Let K be a field and G a finite group. For the following group algebras decide whether they are semisimple or separable:*

(i) $K = GF(2)$, $G = Q_8$

(ii) $K = GF(2)$, G an Abelian 3-group

(iii) $K = \mathbb{R}$, $G = Z_4$

(iv) $K = \mathbb{Q}$, $G = A_6$

(v) $K = GF(3)$, $G = Q_8$.

Excercise 104 *Within counterexample 1 determine all semisimple and separable subalgebras. What is the maximal dimension of all semisimple resp. separable subalgebras? Does a general result exist related to the latter question?*

Excercise 105 *Within example 1 determine all nilpotent subalgebras and all nilpotent left and right ideals. Are they contained in the nilradical? Does a general result exist for the latter question?*

Excercise 106 *Let K be a field and G a finite group. Prove that KG possess an involution. Within the case $K = GF(p)$ (p a prime number) and G is possessing an Abelian normal Hall p-subgroup prove that a radical complement of KG is invariant under this involution.*

Excercise 107 *Let K be a field and $n \in \mathbb{N}$. Prove that $\Delta_{u,n}$ and $\Delta_{o,n}$ are possessing an involution. Prove that a radical complement of for each algebra exist which is invariant under the corresponding involution.*

Excercise 108 *Within the article [27] analyze how resp. why the theorem of Wedderburn-Malcev is resp. can be used (especially within Hilfssatz 2.1).*

Excercise 109 *Analyze exercise 4 within [8], chapter 6, page 115: Let F be a field, $char(F) = 2$, $K := F(s)$ the field of fractions over F, $A := K[t]/(t^4 - s^2)$. Find $R := rad(A)$ and A/R. Verify that A has no subalgebra isomorphic to A/R. Construct a similar example for a field of arbitrary characteristic $p > 0$.*

Excercise 110 *Analyze exercise 5 within [8], chapter 6, page 115: Let F be a field, $char(F) = 2$, $K := F(s)$ the field of fractions over F, $L := K[t]/(t^2 - s)$ and A be the L with basis $\{1, r\}$, $r^2 = 0$. Considering A as a K-algebra establish that $A/rad(A)$ is isomorphic to L and find two distinct subalgebras of A isomorphic to L (since A is commutative, these subalgebras are not conjugated in A). Construct a similar example for a field of arbitrary characteristic $p > 0$.*

Chapter 2

Non-unitary algebras

Within this chapter we face several topics. We transfer the result of Wedderburn-Malcev to non-unitary algebras. For this, we use the result for unitary algebras and the so-called adjunction of an unit. Another main tool is a insight about unit elements of algebras with zero radicals. For generalizing and transferring the conjugacy part of the Wedderburn-Malcev theorem we use the star-group and analyze the connection of the star composition, star conjugation and the composition within the adjunction of an unit. We are able to transfer the result of Wedderburn-Malcev to non-unitary algebras: radical complements exist and they are all conjugated under the action of the star-group. Radical complements are maximal separable subalgebras and every separable algebra is contained in one specific radical complement up to star-conjugation. As a consequence all isomorphism types of separable subalgebras are already contained in one radical complement. The stabilizer of this group action is determined: it is the centralizer of a radical complement under the nilradical action.

As presented in chapter 1 we focus afterwards on Taft's theorem about G-invariant radical complements. Again by using the adjunction of an unit and the star composition we transfer the result to non-unitary algebras: G-invariant radical complements exists and all G-invariant radical complements are G-orthogonal conjugated based on the star-composition.

The idea of compatability is used to derive the radical complements for ideals and factor algebras based on intersection and factorization.

Finally we discuss an algorithm for calculating a radical complement. The case of a zero radical is solved by solving a system of linear equation. This is presented also within [8]. Afterwards an induction argument is used to derive an algorithm for calculating a basis for a radical complement. If we have calculated a radical complement, then we present the calculation for deriving another radical complement based on the star composition. We can use this radical complement to calculate the decomposition for an element as the sum of a radical complement and an element of the complement. We call such procedure a top-down calculation. The decomposition for an element based on another radical complement is also presented based on a transfer rule.

2.1 Adjunction of an unit

Theorem 12 *(adjunction of an unit) Let A be a K-algebra. The K-space $K \times A$ is – equipped with the multiplication*

$$(c;x)(d;y) := (cd; cy + dx + xy)$$

– a K-algebra with unit element $(1_K; 0_A)$. We denote this algebra by (K, A). The function

$$\varphi : A \longrightarrow K \times A, \ a \longmapsto (0_K; a)$$

is an algebra-monomorphism of A in $K \times A$. By using the so-called structure transport and principle of detoxification we create a K-algebra A^K containing A as K-subalgebra. (K, A) is $\mathcal{A}_1$-isomorphic to A^K.

Proof. see [29] where the structure transport and principle of detoxification are explained, too.⋄

Within the next remark we present some elementary properties of the K-algebra (K, A).

Remark 7 *Let A be a K-algebra and φ as within theorem 12. The following statements are true:*

(i) $A \cong_{\mathcal{A}} A\varphi$

(ii) A is associative resp. commutative if and only if (K, A) is associative resp. commutative.

(iii) Subalgebras, left ideals, right ideals and ideals of A are mapped by φ to corresponding substructure of (K, A). In particular, $A\varphi$ is an ideal of (K, A).

(iv) Let K be a field. A is finite-dimensional if and only if (K, A) is finite-dimensional. If A is finite-dimensional, then $dim_K((K, A)) = dim_K(A) + 1$ is true.

(v) $E((K, A)) \cap A\varphi = \emptyset$

Non-unitary algebras

Proof. ad(i): This statement is a direct consequence of theorem 12.

ad(ii): One implication can be deduced from (i). Let $c, d, e \in K$ and $x, y, z \in A$. If A is commutative, then

$$(c; x)(d; y) = (cd; cy + dx + xy) = (dc; dx + cy + yx) = (d; y)(c; x)$$

is true. Thus, (K, A) is commutative. If A is associative, then

$$\begin{aligned}
((c; x)(d; y))(e; z) &= \\
(cd; cy + dx + xy)(e; z) &= \\
((cd)e; (cd)z + e(cy + dx + xy) + (cy + dx + xy)z) &= \\
(cde; (cd)z + (ec)y + (ed)x + e(xy) + c(yz) + d(xz) + xyz) &\quad \text{and} \\
(c; x)((d; y)(e; z)) &= \\
(c; x)(de; dz + ey + yz) &= \\
(cde; c(dz + ey + yz) + (de)x + x(dz + ey + yz)) &= \\
(cde; (cd)z + (ce)y + c(yz) + (de)x + d(xz) + e(xy) + xyz)
\end{aligned}$$

are true. Thus, by comparing these statements also item (ii) is valid.

ad(iii): By using statement (i) we conclude that the φ-images of the mentioned structures are subalgebras of (K, A). For all $k \in K$ and $t, a \in A$ the equations $(0_K; t)(k; a) = (0_K; kt + ta)$ and $(k; a)(0_K; t) = (0_K; kt + at)$ are valid, and thus (iii) is true.

ad(iv): This is a consequence of basic linear algebraic facts.

ad(v): $A\varphi \neq (K, A)$ is valid, and thus (v) can be deduced from (iii). ◇

Within the next lemma we illustrate additional properties of the K-algebra (K, A). Items (i) and (ii) of this lemma will play an important role for determining the nilradical of (K, A), the last item is used for the conjugacy theorem for non-unital algebras.

Lemma 2 *Let A be a K-algebra and φ as used within theorem 12. The following statements are valid:*

(i) If A is unitary, then $(K, A) \cong_{A_1} K \oplus A$ is true.
Let $J := \langle (1_K; -1_A) \rangle_K$. $\{J, A\varphi\}$ is a direct decomposition into ideals of (K, A) and, in addition, $A \cong_{A_1} A\varphi$ and $K \cong_{A_1} J$ are true.

(ii) If I is an ideal of A, then $(K, A)/I\varphi \cong_{A_1} (K, A/I)$ is true.

(iii) Let T be a K-subalgebra of A. $T\varphi$ and (K,T) are K-subalgebras of (K,A). (K,T) is unital.[1]

Proof. ad(i): We define
$$\alpha : (K,A) \longrightarrow K \oplus A, \ (k;a) \longmapsto (k; k1_A + a).$$
α is an algebra isomorphism which is proven within the following lines. Let $k, k', l \in K$, $a, a' \in A$. The identity
$$((k;a) + (k';a'))\alpha =$$
$$(k + k'; (k + k')1_A + (a + a')) =$$
$$(k;a)\alpha + (k';a')\alpha$$
is true. In addition, the statements $(l(k;a))\alpha = (lk; lk1_A + la) = l((k;a)\alpha)$, $\ker \alpha = \{(0_K; 0_A)\}$ and $(k; a - k1_A)\alpha = (k;a)$ are true. Hence, α is a K-space isomorphism. The unit element is mapped to the unit element: $(1_K; 0_A)\alpha = (1_K; 1_A)$ is true. A straightforward calculation shows us
$$((k;a)(k';a'))\alpha =$$
$$(kk'; kk'1_A + ka' + k'a + aa') =$$
$$(k;a)\alpha(k';a')\alpha.$$
For all $(k;a) \in K \times A$ the identity $(k;a)\alpha^{-1} = (k; -k1_A + a)$ is valid. Hence, item (i) is proven.

ad(ii): Straightforward to calculate is that the function
$$\beta : (K,A) \longrightarrow (K, A/I), \ (k;a) \longmapsto (k; a + I)$$
is a $\mathcal{A}_1$-epimorphism possessing the kernel $\ker \beta = I\varphi$.

ad(iii): Item (iii) is a consequence of theorem 12.⋄

Note 2 Within this remark we prove that the condition (i) of lemma 2 for an algebra based on a field is equivalent of being unitary. Let K be a field, A a K-algebra, $\{I, J\}$ a direct decomposition of (K,A) into ideals of (K,A), such that $I \cong_A K$ and $J \cong_A A$ are valid. K is an unitary K-algebra. Hence, I is unitary, too. By using theorem 12 we deduce that (K,A) is an unitary K-algebra. Thus, $j \in J$ and $k \in K$ exist such that $1_{(K,A)} = k1_I + j$ is valid. We prove that j is the unit element of J. Let $x \in J$. The statement
$$x = x1_{(K,A)} = k1_I x + jx = jx$$
is true and $x = jx$ is valid. By using an analogue calculation we prove $x = xj$. Hence, j is the unit element of J. A and J are isomorphic, and thus A is unitary, too.⋄

---

[1] Unital subalgebras contain the unit element of the algebra they are contained in. Unitary subalgebras are unital as algebras. They are not unital in general.

2.2 The existence part

The next lemma is another important fact used for determining the nilradical of (K, A). We will prove that every finite-dimensional associative semisimple K-algebra is unitary. This unit is used later on to transfer the unitary version of the theorem of Wedderburn-Malcev to non-unitary algebras.

Lemma 1 *(unitarity of semisimple algebras) Every finite-dimensional associative semisimple K-algebra is unitary. In particular, the factor algebra by the nilradical of a finite-dimensional associative K-algebra is unitary.*

Proof. (see [29]) This proof is divided into two parts.

(i) We begin the proof by showing that every left unit is an unit of A. Let e be a left unit of A. We define $B := \{x \mid x \in A, xe = 0_A\}$. It is straightforward to prove that B is a K-subspace of A. Let $x \in B$ and $a \in A$. We calculate

$$xa = xea = 0_A a = 0_A.$$

Hence, B is a zero K-right ideal of A contained in $rad(A)$. Therefor, $B = \{0_A\}$ is valid. Let $a \in A$. The condition $a - ae \in B$ is valid, and thus we deduce $a - ae = 0_A$. Hence $a = ae$ is valid and the proof of (i) is finished.

(ii) Now we will construct a left unit for A. A is finite-dimensional and associative, and thus idempotent elements $e_1, ..., e_k$ of A and a K-right ideal S of A exist such that

$$A = e_1 A \oplus_K ... \oplus_K e_k A \oplus_K S$$

is valid. In addition, for all $i, j \in \underline{k}$ for which $i < j$ is valid the statements $e_i e_j = 0_A$ is true and S possesses no idempotent $\neq 0_A$. S is a nilpotent K-right ideal of A which is contained in $rad(A)$. A is semisimple and therefor $S = \{0_A\}$ is valid. For all $r \in \underline{k}$ we define

$$s_r := (-1_K)^{r-1} \sum_{i_1 > ... > i_r} e_{i_1} ... e_{i_r},$$

and we will prove that $e := \sum_{r=1}^{k} s_r$ is a left unit of A. For this we have to deduce that for all $t \in \underline{k}$ the condition $ee_t = e_t$ is true. Let $r, t \in \underline{k}$. Straightforward to prove is the equation

$$s_r e_t = (-1_K)^{r-1} \Big(\sum_{i_1 > ... i_r > t} e_{i_1} ... e_{i_r} e_t + \sum_{i_1 > ... > i_r = t} e_{i_1} ... e_{i_r} \Big).$$

A telescoping sum argument implies

$$ee_t = e_t + (-1_K)^{k-1} \sum_{i_1 > \ldots i_k > t} e_{i_1}\ldots e_{i_k} e_t = e_t.$$

Thus, the theorem is proven.◇

By using this result we are almost ready to determine the nilradical and the factor algebra by the nilradical of (K, A). Within the next proposition we present some properties of separable algebras which are needed for this analysis.

Proposition 2 *Let K be a field, A, B associative K-algebras and I an ideal of A.*

(i) Let A be separable. I and A/I are separable. K-subalgebras of A are not separable in general.

(ii) A and B are separable if and only if $A \oplus B$ is separable.

(iii) If A/I and I are separable, then A is separable.

Proof. ad(i): Based on A also the ideal I of A is – by using part (i) of corollary 1 – finite-dimensional and semisimple. Lemma 1 lets us deduce that I is unitary. In addition, $I^- \otimes_K I$ is semisimple as K-ideal of the semisimple K-algebra $A^- \otimes_K A$ (see statement (ii) of theorem 1). Hence, I is – based on statement (ii) of theorem 1 – separable. A is finite-dimensional and unitary and so is A/I. We have to prove (see part (ii) of theorem 1) that $(A/I)^- \otimes_K A/I$ is semisimple. Let π be the canonical algebra epimorphism from A onto A/I. It is straightforward to show that π is an algebra epimorphism from A^- onto $(A/I)^-$. Thus, $\pi \otimes \pi$ is an algebra epimorphism from $A^- \otimes_K A$ onto $(A/I)^- \otimes_K A/I$. The latter algebra is semisimple as image of the semisimple K-algebra $A^- \otimes_K A$ under $\pi \otimes \pi$.
Let us focus on the separable K-algebra $K^{n \times n}$ for $n \neq 1$. The subalgebra of lower triangular matrices is not semisimple (see also example iv, the section 1.3.2 for lower triangular matrices and part (i) of corollary 1).

ad(ii): If $A \oplus B$ is separable, then A and B are separable based on part (i). The opposite statement is true based on part (iv) of theorem 1.

ad(iii): Based on part (i) of corollary 1 the K-algebras A/I and I are finite-dimensional and semisimple. By using $rad(A/I) = (rad(A)+I)/I = \{I\}$ we deduce $rad(A) \subseteq I$. Hence, $rad(A) = rad(A) \cap I = rad(I) = \{0_A\}$ is valid. This implies that A is semisimple. As a consequence the K-ideal I possesses an ideal complement in A. This complement – which is isomorphic to A/I A_1 – is separable. Part (ii) finishes the proof.◇

Non-unitary algebras   63

Corollary 3 *(the nilradical and its factor algebra within an adjunction of an unit)* Let K be a field, A a finite-dimensional associative K-algebra and φ as defined within theorem 12. The following statements are true:

(i) (K, A) is an associative finite-dimensional unitary K-algebra.

(ii) $rad((K, A)) = rad(A)\varphi$

(iii) $(K, A)/rad((K, A)) \cong_{A_1} K \oplus A/rad(A)$

(iv) $(K, A)/rad((K, A))$ is separable if and only if $A/rad(A)$ is separable.

Proof. ad(i): This statement is a direct consequence of theorem 12 and of the statements (ii) and (iv) of remark 7.

ad(ii) and (iii): By using theorem 12 and part (i) of remark 7 we deduce that $rad(A)\varphi$ is a nilpotent ideal of the associative finite-dimensional unitary K-algebra (K, A) (see also part (i)). We apply statement (ii) of lemma 2 and deduce $(K, A)/rad(A)\varphi \cong_{A_1} (K, A/rad(A))$. Theorem 1 is used to prove that $A/rad(A)$ is unitary. Thus, by using statement (i) of lemma 2 we deduce $(K, A)/rad(A)\varphi \cong_{A_1} K \oplus A/rad(A)$. Direct products of semisimple algebras are semisimple. Hence, the statements (ii) and (iii) are proven.

ad(iv): K is separable as K-algebra. Hence, statement (iv) is a consequence of statement (ii) and part (ii) of proposition 2.⋄

The following Hasse-diagram illustrates the proven results so far about the algebra (K, A):

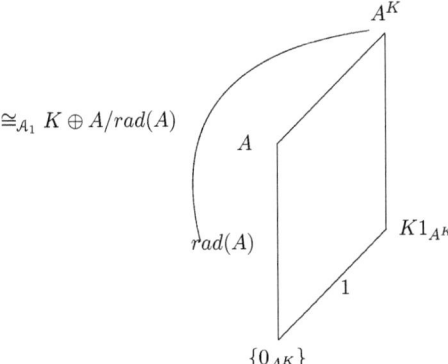

Now we prove the existence of a radical complement for non-unitary algebras:

Theorem 13 *(existence of radical complements for non-unitary algebras)*
Let K be a field and A a finite-dimensional associative K-algebra. If $A/rad(A)$ is separable, then $rad(A)$ possesses an algebra complement in A.

Proof. Let φ be as defined within theorem 12. By using theorem 12 and corollary 3 we deduce that (K, A) is a finite-dimensional associative unitary K-algebra such that $rad((K, A)) = rad(A\varphi)$ is valid and $(K, A)/rad((K, A))$ is separable. Now we apply the theorem of Wedderburn-Malcev (theorem 8) and deduce that $rad((K, A))$ possesses an algebra complement T in (K, A). By using Dedekind's identity we deduce

$$A\varphi = (K, A) \cap A\varphi = (rad(A)\varphi \oplus_K T) \cap A\varphi = rad(A)\varphi \oplus_K (T \cap A\varphi).$$

Statement (i) of remark 7 finishes the proof.⋄

We end this section by applying the proven theorem 13 to the tensor product of two algebras for determining its nilradical and the factor algebra by its nilradical. The next three lemmas are needed for this analysis.

Lemma 3 *Let A be a K-algebra, I a K-ideal of A and S a K-ideal of the K-subalgebra T of A. The following statements are valid:*

(i) $I + S$ is a K-ideal of $I + T$.

(ii) $(I \cap T) + S$ is a K-ideal of T.

(iii) The K-algebras $(I + T)/(I + S)$ and $T/((I \cap T) + S)$ are isomorphic.

Proof. ad(i) and (ii): The statements (i) and (ii) are straightforward to prove.

ad(iii): By using Dedekind's identity be deduce $(I + S) \cap T = (I \cap T) + S$. Thus, the statement (iii) is a consequence of basic isomorphism theorems for associative algebras.⋄

Lemma 4 *Let K be a field, A, B finite-dimensional associative K-algebras and I a K-ideal of B. $A \otimes_K I$ is a K-ideal of $A \otimes_K B$ and the statement $(A \otimes_K B)/(A \otimes_K I) \cong_A A \otimes_K (B/I)$ is valid.*

Proof. Let γ be the function from $A \otimes_K B$ to $A \otimes_K (B/I)$ defined by $(a \otimes b)\gamma = a \otimes (b + I)$ for all $(a; b) \in A \times B$. It is straightforward to prove that γ is a A-epimorphism for which the statement $A \otimes_K I \subseteq ker\gamma$ is valid. By using basic isomorphism theorems for associative algebras we deduce $dim_K(ker\gamma) = dim_K(A \otimes_K I)$. Hence, $ker\gamma = A \otimes_K I$ is valid and the proof is finished.⋄

Lemma 5 *(separable algebras and tensor products) Let K be a field, A an associative separable K-algebra and B a finite-dimensional associative semisimple K-algebra. The K-algebra $A \otimes_K B$ is semisimple.*

Proof. see theorem 71.10 in [6] or execute exercise 287. ⋄

The following examples are related to lemma 5.

Example 3 (i) In the context of finite-dimensional central-simple associative K-algebras the statements $\mathbb{H} \otimes_{\mathbb{R}} \mathbb{H}^- \cong_{A_1} \mathbb{R}^{4 \times 4}$ and $\mathbb{H} \otimes_{\mathbb{R}} \mathbb{C} \cong_{A_1} \mathbb{C}^{2 \times 2}$ are proven. Hence, the tensor product of two K-division algebras need not to be a K-division algebra, too.

(ii) A straightforward calculation shows us that $\mathbb{C} \otimes_{\mathbb{R}} \mathbb{C} \cong_{A_1} \mathbb{C} \oplus \mathbb{C}$ is valid. Hence, the tensor product of two simple K-algebras need not to be simple, too.

(iii) Let $(K; L)$ be a finite-dimensional, non-separable field extension (see e.g. example 1). By using theorem 1 we deduce that the K-algebra $L \otimes_K L$ is not semisimple. Hence, the tensor product of two semisimple K-algebras need not to be semisimple, too. ⋄

Theorem 14 *Let K be a field and A, B finite-dimensional associative K-algebras. The following statements are valid:*

(i) *If $A/rad(A)$ or $B/rad(B)$ is separable, then the identities $rad(A \otimes_K B) = rad(A) \otimes_K B + A \otimes_K rad(B)$ and $(A \otimes_K B)/rad(A \otimes_K B) \cong_{A_1} (A/rad(A)) \otimes_K (B/rad(B))$ are true.*

(ii) *If $A/rad(A)$ and $B/rad(B)$ are separable and S resp. T is a radical complement in A resp. B, then $S \otimes_K T$ is a radical complement in $A \otimes_K B$ and $(A \otimes_K B)/rad(A \otimes_K B)$ is separable.*

Proof. ad(i): Because of $A \otimes_K B \cong_A B \otimes_K A$ we assume that $A/rad(A)$ is separable. Let $I := rad(A) \otimes_K B + A \otimes_K rad(B)$. It is straightforward to prove that I is a nilpotent ideal of $A \otimes_K B$. Thus, we have only to prove that $(A \otimes_K B)/I$ is semisimple. $A/rad(A)$ is separable. Hence, by using theorem 13 a radical complement T in A exists. We deduce $A \otimes_K B = rad(A) \otimes_K B + T \otimes_K B$ and $I = rad(A) \otimes_K B + T \otimes_K rad(B)$. Lemma 3 lets us deduce that $(A \otimes_K B)/I \cong_A ((rad(A) \otimes_K B) \cap (T \otimes_K B)) + T \otimes_K rad(B)$ is valid. Using $rad(A) \oplus_K T = A$ and lemma 4 we conclude $(A \otimes_K B)/I \cong_A T \otimes_K (B/rad(B))$. Lemma 5 implies statement (i).

ad(ii): Statement (ii) is deductable from theorem 13, statement (i) and lemma 5. ⋄

We apply this theorem within the next example.

Example 4 *($\otimes = \oplus$)* Within part (ii) of example 3 we have remarked that $\mathbb{C} \otimes_\mathbb{R} \mathbb{C} \cong_{A_1} \mathbb{C} \oplus \mathbb{C}$ is true. Based on this statement we analyze the question for which finite-dimensional associative unitary K-algebras A the statement $A \otimes_K A \cong_{A_1} A \oplus A$ is valid. Straightforward to prove is that the condition $dim_K(A) = 2$ is true (The zero-space is not possible because of $1 \neq 0$.). Theorem 1.1.1 in [8] classifies K-algebras of K-dimension 2.

Case 1:
In this case $A \cong_{A_1} K^2$ is assumed. The initial question is answered positively because the statement $K^2 \otimes K^2 \cong_{A_1} K^4 \cong_{A_1} K^2 \times K^2$ is true.

Case 2:
Within this case A possess a one-dimensional nilradical. Hence, $dim_K(rad(A \oplus A)) = 2$ is valid. By using theorem 14 we deduce $dim_K(rad(A \otimes_K A)) = 1$, and thus the initial question is answered negatively.

Case 3:
In this case we face a two-dimensional field extension $(K1_A; A)$. Let us assume that the field extension is inseparable. The K-algebra $A \oplus A$ is semisimple. By using theorem 1 and corollary 1 we deduce that the K-algebra $A \otimes_K A$ is not semisimple. In this subcase the initial question is answered negatively.
Now we assume that $(K1_A; A)$ is separable. Hence, the field extension is a Galois extension of degree 2. Thus, an element $a \in A$ exists such that $A = K1_A(a)$ is valid. Based on proposition 5.3.1 in [8] we deduce that $A \otimes_K A$ and $K1_A(a)[t]/(K1_A(a)[t]f)$ are isomorphic. Within this statement f is the minimal polynomial of a over $K1_A$. f is separable over $K1_A$ ist, and thus f decomposes in $K1_A(a)[t]$ in two different linear factors. By using the Chinese Remainder theorem we deduce that the initial question is answered positively.◊

2.3 The conjugacy part

Within this section we focus on non-unitary associative algebras. We have to answer the question in what way the conjugacy part of the Wedderburn-Malcev theorem can be formulated. For this proposition 3 is relevant. Within part (v) of proposition 3 we define a well-known generalization of the group of units which we will use frequently within this work.

Definition 4 *(shift)* Let A be an Abelian group and $a \in A$. We define
$$\alpha_a : A \longrightarrow A, x \longmapsto x + a.$$

This function is called the shift by a in A.⋄

Proposition 3 *(star composition) Let A be a K-algebra. For all $a, b \in A$ let $a * b := a + b + ab$. For the star or circle composition $*$ the following statements are valid:*

(i) *$(A; *)$ is a monoid possessing the unit element 0_A. If $(A; *)$ is associative, then $Q(A)$ denotes the group of units of this monoid. This group is called the star or quasi regular group. By $a^{(-1)}$ or a' we denote the inverse of $a \in Q(A)$.*

(ii) *For all $a, b, c \in A$ the condition $ab = ba$ resp. $(ab)c = a(bc)$ is valid if and only if $a * b = b * a$ resp. $(a * b) * c = a * (b * c)$ is true. In particular, $(A; \cdot)$ is commutative resp. associative if and only if $(A; *)$ is commutative resp. associative. If $(A; \cdot)$ is associative, then for all $a \in A$ and $r \in Q(A)$ the statement*

$$r^{(-1)} * a * r = a + r^{(-1)}a + ar + r^{(-1)}ar$$

is valid.

(iii) *If A is unitary, then α_{1_A} is a monoid isomorphisms between $(A; *)$ and $(A; \cdot)$.*

(iv) *If $\gamma : A \longrightarrow B$ is an algebra homomorphism, then γ is a magma homomorphism between $(A; *)$ and $(B; *)$.*

(v) *Let A be associative.*

   (a) *If a is a nilpotent element of A, then $a \in Q(A)$ and*

   $$a^{(-1)} = \sum_{s=1}^{cl(a)-1} (-1_K)^s a^s$$

   are valid. In particular, $(rad(A); *)$ is a group.

   (b) *$rad(A)$ is – if A is right artian – a nilpotent normal subgroup of $Q(A)$.*

   (c) *If A is unitary and right artian, then α_{1_A} restricted to $Q(A)$ is a group isomorphism to $E(A)$. $rad(A)\alpha_{1_A} = 1_A + rad(A)$ is a (nilpotent) normal subgroup of $E(A)$.*

Proof. see e.g. [29].⋄

Within the next remark the connection between the adjunction of an unit and the star composition is analyzed.

Remark 8 Let A be a K-algebra, S a K-subspace, T a K-subalgebra of A and φ as defined in theorem 12. The following statements are valid:

(i) If $A = S \oplus_K T$ is valid, then $(K, A) = S\varphi \oplus_K (K, T)$ is true.

(ii) For all $k, l \in K$ and $a, b \in A$ the identity $(k; a) * (l; b) = (k * l; a * b + kb + la)$ is valid.

(iii) For all $a, c \in A, k \in K$ the identity $a\varphi * (k; c) \in A\varphi$ resp. $(k; c) * a\varphi \in A\varphi$ is true if and only if $k = 0_K$ is valid.

(iv) If C is a K-subalgebra of A, then for all $a \in A$ the statements $(a * C)\varphi = a\varphi * C\varphi = (a\varphi * (K, C)) \cap A\varphi$ and $(C * a)\varphi = C\varphi * a\varphi = ((K, C) * a\varphi) \cap A\varphi$ are valid.$\diamond$

Definition 5 *(conjugation within the circle composition)* Let A be a associative K-algebra and $r \in Q(A)$. We define

$$\kappa_{(r)} : A \longrightarrow A, a \longmapsto r^{(-1)} * a * r.$$

For all $a \in A$ and $T \subseteq A$ we use instead of $a\kappa_{(r)}$ resp. $T\kappa_{(r)}$ the symbols $a^{(r)}$ resp. $T^{(r)}$ and call $\kappa_{(r)}$ the conjugation with r.$\diamond$

Theorem 15 *(conjugacy of radical complements)* Let K be a field and A a finite-dimensional associative K-algebra possessing a separable factor algebra by its nilradical. If S, T radical complements in A, then an element $r \in rad(A)$ exists such that $T = S^{(r)}$ is valid.

Proof. Let φ as defined within theorem 12. By using corollary 3 the pair (K, A) is a finite-dimensional associative unitary K-algebra possessing the nilradical $rad((K, A)) = rad(A\varphi)$ and a separable factor algebra by its nilradical. Part (iii) of lemma 2 and part (i) of remark 8 let us deduce that (K, S) and (K, T) are radical complements. Based on theorem 8 an element $r \in rad(A)$ exists such that $(K, S) = (1_K; r)^{-1}(K, T)(1_K; r)$ is valid. By using part (iii) of remark 3 and remark 4 we conclude that $(K, S) = (0_K; r)^{(-1)} * (K, T) * (0_K; r)$ and $r\varphi * (K, S) = (K, T) * r\varphi$ are valid. An intersection with $A\varphi$ and using part (iv) of remark 8 result in $(r * S)\varphi = (T * r)\varphi$. φ is injective, and thus $r * S = T * r$ and $T = S^{(r)}$ are proven.$\diamond$

A similarity to the theorem of Sylow is that each separable subalgebra can be conjugated into one specific radical complement. Thus, every isomorphic class of separable subalgebras are contained in one specific radical complement. For that proof we need the following properties of conjugation based on the star composition.

Remark 9 Let A be an associative K-algebra and $x \in Q(A)$. $\kappa_{(x)}$ is an algebra automorphism of A. In particular, the function

$$\kappa_{()} : Q(A) \longrightarrow Aut_K(A), x \longmapsto \kappa_{(x)}$$

is a group homomorphism possessing the kernel $\ker \kappa_{()} = Z(A) \cap Q(A)$, and the image $\kappa_{()}$ is a normal subgroup of $Aut_K(A)$. For all $x \in Q(A)$ and $\alpha \in Aut_K(A)$ the statement $\kappa_{(x)}{}^\alpha = \kappa_{(x\alpha)}$ is valid.

Proof. It is well-known that $\kappa_{(x)}$ is a monoid automorphism of $(A; *)$. Hence, the function is bijective. By using part (ii) of remark 3 for all $a \in A$ the statement

$$(*)\ x^{(-1)} * a * x = a + x^{(-1)}a + ax + x^{(-1)}ax$$

is valid.
Let $a, b \in A, k \in K$. Based on $(*)$ the identity

$$x^{(-1)} * (ka) * x = ka + x^{(-1)}(ka) + (ka)x + x^{(-1)}(ka)x = k(x^{(-1)} * a * x)$$

is true.
Again by using $(*)$ the statement

$$x^{(-1)} * (a+b) * x =$$
$$(a+b) + x^{(-1)}(a+b) + (a+b)x + x^{(-1)}(a+b)x =$$
$$x^{(-1)} * a * x + x^{(-1)} * b * x$$

is valid.
Another usage of $(*)$ implies

$$(x^{(-1)} * a * x)(x^{(-1)} * b * x) =$$
$$ab + ax^{(-1)}b + abx + ax^{(-1)}bx +$$
$$x^{(-1)}ab + x^{(-1)}ax^{(-1)}b + x^{(-1)}abx + x^{(-1)}ax^{(-1)}bx +$$
$$+axb + axx^{(-1)}b + axbx + axx^{(-1)}bx + x^{(-1)}axb +$$
$$x^{(-1)}axx^{(-1)}b + x^{(-1)}axbx + x^{(-1)}axx^{(-1)}bx.$$

Because of $x \in Q(A)$ we conclude

$$ax^{(-1)}b + axb + axx^{(-1)}b = 0_A,$$
$$x^{(-1)}ax^{(-1)}bx + x^{(-1)}axbx + x^{(-1)}axx^{(-1)}bx = 0_A,$$
$$axbx + ax^{(-1)}bx + axx^{(-1)}bx = 0_A \text{ and}$$
$$x^{(-1)}ax^{(-1)}b + x^{(-1)}axb + x^{(-1)}axx^{(-1)}b = 0_A.$$

Hence,

$$(x^{(-1)} * a * x)(x^{(-1)} * b * x) = ab + abx + x^{(-1)}ab + x^{(-1)}abx$$

is valid. Again by using (∗) the statement for $\kappa_{(x)}$ is proven.
$Q(A) = E((A; *))$ is valid, and thus $\kappa_{()}$ is a group homomorphism of $Q(A)$ into $Aut((A; *))$. It is straightforward to prove that its kernel is exactly $Z(A) \cap Q(A)$. Based on part (iv) and theorem 3 we deduce that $Aut_K(A) \subseteq Aut((A; *))$ is valid. Thus, the last statement within the remark is true for all $\alpha \in Aut((A; *))$. ⋄

Corollary 4 *(maximality of radical complements) Let K be a field and A a finite-dimensional associative K-algebra possessing a separable factor algebra by its nilradical. The following statements are valid:*

(i) *If T is a radical complement of A and S is a separable K-subalgebra of A, then an element $r \in rad(A)$ exists such that $S^{(r)} \subseteq T$ is valid.*

(ii) *$(rad(A); \star)$ acts transitive by conjugation on the set of all radical complements of A.*

(iii) *The radical complements of A are exactly the separable K-subalgebras of maximal K-dimension of A.*

(iv) *The radical complements of A are exactly the maximal elements with respect to $\subseteq$ of the set of separable K-subalgebras of A.*

Proof. ad(i): Let $B := rad(A) + S$. $S \cap rad(A)$ is a nilpotent K-ideal of S, and thus $B = rad(A) \oplus_K S$ is valid. In addition, $rad(B) = rad(A)$ is true. B is a finite-dimensional associative K-algebra. We prove that $T \cap B$ is a radical complement of B. By using $rad(A) \subseteq B$ and Dedekind's law we conclude

$$rad(A) \oplus_K (T \cap B) = (rad(A) \oplus_K T) \cap B = A \cap B = B.$$

Hence, $T \cap B$ possesses the required property. Based on theorem 15 an element $r \in rad(A)$ exists such that $S^{(r)} = T \cap B \subseteq T$ is valid. Thus, part (i) is proven.

ad(ii): This statement is a direct consequence of theorem 15 and remark 9.

ad(iii): Let T be a radical complement of A existing based on theorem 13. By using part (i) and remark 9 we deduce that T has the required properties. Let S be a separable K-subalgebra of A possessing maximal K-dimension within all separable K-subalgebras. By using part (i), remark 9 and the statement just proven an element $r \in rad(A)$ exists such that $S^{(r)} = T$ is valid. Part (iii) is now proven using part (ii).

ad(iv): Let T be a radical complement of A existing based on theorem

13. If M is a maximal element of the set of separable K-subalgebras of A, then – by using part (i) – an element $r \in rad(A)$ exists such that $M^{(r)} \subseteq T$ is valid. $M^{(r)}$ is a maximal element of the same set. Hence, $M^{(r)} = T$ is valid. By using part (ii) the K-subalgebra $M = T^{(r^{(-1)})}$ is a radical complement of A. The implication $\Longrightarrow$ is a consequence of part (iii).◇

We finish this section by analyzing an example of a non-unitary algebra and by transferring the results proven so far from the non-unitary case to the unitary one.

Example 5 Let K be a field and A a 2-dimensional K-algebra possessing a K-basis $\{e, r\}$ equipped with the multiplication

$\cdot$	e	r
e	e	r
r	0_A	0_A

A is a non-commutative associative K-algebra which is $\mathcal{A}$-isomorphic to a K-subalgebra of the K-algebra within example 1. It is straightforward to prove that $rad(A) = \langle r \rangle_K$ is valid, that $T := \langle e \rangle_K$ is a radical complement and that the factor algebra by the nilradical is separable.

Now we prove that A is non-unitary. For this, we assume that A is unitary. In this case elements $k, l \in K$ would exist such that $1_A = ke + lr$ is valid. Because of

$$e = (ke + lr)e = ke$$

we would deduce $k = 1_K$, and by using

$$e = e(e + lr) = e + lr$$

we would prove $l = 0_K$. Thus, $1_A = e$ would be valid which would imply $r = re = 0_A$. This is a contradiction. Hence, A is non-unitary.

By using theorem 15 we are still able to determine all radical complements in A. Let $k \in K$. The identity $(-kr) * (kr) = 0_A$ is true, and thus

$$e^{(kr)} = (-kr) * e * (kr) = (-kr + e) * kr = e + kr$$

is valid. Hence, $\{\langle e + kr \rangle_K \mid r \in rad(A)\}$ is the set of all radical complements of A. If $k, l \in K$ such that $\langle e + kr \rangle_K = \langle e + lr \rangle_K$ is true, then $k = l$ is valid. Hence, there is a bijection from $rad(A)$ onto the set of all radical complements of A.

We conclude this example by determining all subalgebras of A. Let S be an one-dimensional subalgebra of A. If $S \cap rad(A) = \{0_A\}$ is valid, then S is (based on a dimension argument) a radical complement. In the other case S must be equal to $rad(A)$. The following picture illustrated the structure of the algebra A.⋄

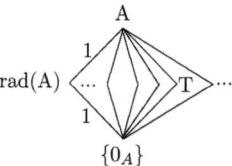

The following remark clarifies the connection between conjugation based on the star composition and based on the ordinary algebra multiplication for an unitary algebra: they are identical.

Remark 10 *Let A be an associative unitary K-algebra and $r \in Q(A)$. The statement $\kappa_{(r)} = \kappa_{1_A + r}$ is valid.*

Proof. Let $a \in A$. Based on part (ii) of remark 3 the identity
$$a^{(r)} = a + r^{(-1)}a + ar + r^{(-1)}ar$$
is valid. By using part (iii) of remark 3 we deduce
$$(1_A + r)^{-1} = 1_A + r^{(-1)}.$$
Hence,
$$a^{1_A + r} = (a + r^{(-1)}a)(1_A + r) = a^{(r)}$$
is proven.⋄

By using this remark we are able to transfer statements proven for nonunitary algebras to unitary algebras, e.g.: If within the context of corollary 4 the algebra is unitary, then each separable subalgebra can be conjugated into one fixed radical complement based on an element $1 + r$, $r \in rad(A)$.

2.4 Cardinality of the set of radical complements

Based on part (ii) of corollary 4 the group $(rad(A); *)$ acts transitive on the set of radical complements of A by conjugation. Remark 9 lets us deduce that this group acts per conjugation also on the complete set of separable K-subalgebras of A. This action is not transitive in general because of

Non-unitary algebras   73

dimension reasons. One possible description of this action could be that the orbits are the sets of separable K-subalgebras of equal dimension which are invariant under the action of $(rad(A); *)$. But this is not true as the next example will show us. The description of the orbits remains as an open topic. But we will prove some aspects of the cardinality of the set of all radical complements. As an example we will focus again on example 1 and on the section 1.3.2 of triangular matrices. In the latter case we have already determined this cardinality.

Counterexample 2 Let K be a field. We focus on (see part (iv) of theorem 1) the associative commutative separable K-algebra $K \times K$. The K-ideals $K \times \{0_K\}$ and $\{0_K\} \times K$ of A – which are isomorphic to K and separable – are distinct. In particular, both are not conjugated in $K \times K$. This example can generalized to any commutative algebra possessing two distinct separable K-subalgebras of the same dimension.◊

We start this section by stating some remarks concerning group actions.

Definitions 4 Let G be a group and U a subgroup of G. By G/U or $G/_r U$ we denote the set of right cosets of U in G. If M is a G-set, then for every $m \in M$ the orbit of m under the action of G resp. the stabilizer of m in G is symbolized by mG resp. by $Stab_G(m)$.◊

Remark 11 Let G be a group and M a G-set. If $m \in M$, then the identity $\mid G/Stab_G(m) \mid = \mid mG \mid$ is valid. The function

$$\psi_m : G \longrightarrow mG, g \longmapsto mg$$

is surjective. It is bijective if and only if $Stab_G(m) = \{1_G\}$ is true.◊

The analysis starts by proving a connection between normalizers and centralizers for semisimple subalgebras. If A is an associative K-algebra and D a subset of A, then we define the normalizer and centralizer of D by $N_A(D) := \{a \in A \mid aD = Da\}$ and $C_A(D) := \{a \in A \mid \forall d \in D : ad = da\}$. If A is a group with respect to $\star$, then we symbolize this by using $A^\star = Q(A)$.

Remark 12 Let A be an associative unitary K-algebra and D a semisimple K-subalgebra of A. The identity

$$N_A(D) \cap (1_A + rad(A)) = C_A(D) \cap (1_A + rad(A))$$

is valid. In particular,

$$N_A(D) \cap rad(A) = C_A(D) \cap rad(A)$$

is true.

Proof. Let $r \in N_A(D) \cap (1_A + rad(A))$. If $d \in D$, then an element $t \in D$ exists such that $(1_A + r)d = t(1_A + r)$ is valid. Hence, $d - t = tr - rd \in rad(A) \cap D$ is true. Because of $rad(A) + D = rad(A) \oplus_K D$ we deduce $d = t$. Thus, $rd = dr$ is valid. The other inclusion is straightforward to prove. The other part of this remark is proven by applying the bijective function α_{-1_A} to the identity just proven.⋄

Theorem 16 *(cardinality of the set of radical complements) Let K be a field, A a finite-dimensional associative K-algebra, D a semisimple K-subalgebra of A and $B := rad(A) \oplus_K D$. The normal subgroup $rad(A)$ of the star group of A acts via conjugation on the set of all semisimple K-subalgebras of A (see remark 9). Based on remark 11 we deduce that*

$$\mid rad(A)^\star / Stab_{rad(A)^\star}(D) \mid = \mid D^{(rad(A))} \mid$$

is valid. Let

$$\psi_D : rad(A) \longmapsto D^{(rad(A))}, r \longmapsto D^{(r)}.$$

The following statements are valid::

(i) $Stab_{rad(A)^\star}(D) = N_A(D) \cap rad(A) = C_A(D) \cap rad(A)$.

(ii) *The function ψ_D is bijective if and only if $C_A(D) \cap rad(A) = \{0_A\}$ is valid.*

In this case we say that D possesses 'maximal orbit length'.

(iii) *If $D \subseteq C_B(D)$ is valid, then ψ_D is bijective if and only if the condition $D = C_B(D)$ is true.*

(iv) *D is commutative if and only if $D \subseteq C_B(D) = N_B(D)$ is valid.*

(v) *Within the examples 1 and 1.3.2 every radical complement possesses maximal orbit length.*

(vi) *$\mid D^{(rad(A))} \mid = 1$ is valid if and only if $rad(A) \subseteq C_B(D)$ is true.*

In this case we say that D possesses 'minimal orbit length'. One sufficient condition for this is that B is commutative.

(vii) *Let $A/rad(A)$ be separable and T a radical complement. T possesses minimal orbit length if every separable K-subalgebra possesses minimal orbit length.*

Proof. ad(i): Let $r \in Stab_{rad(A)^\star}(D)$ and $d \in D$. Because of $D * r = r * D$ an element $t \in D$ exists such that $d + dr = rt + t$ is valid. Based on $rad(A) + D = rad(A) \oplus_K D$ we deduce $d = t$ and $rd = dr$. Thus,

Non-unitary algebras   75

$r \in C_A(D) \cap rad(A)$ is proven. Let $r \in C_A(D) \cap rad(A)$. Hence, $dr = rd$ is valid for all $d \in D$. Based on part (ii) of theorem 3 we deduce $D * r = r * D$, and this implies $r \in Stab_{rad(A)}(D)$. We only have to prove $N_A(D) \cap rad(A) \subseteq C_A(D) \cap rad(A)$. Let $r \in N_A(D) \cap rad(A)$. Within the K-algebra (K, A) for all $k \in K$ and $d \in D$ the equations $(k;d)(0_K;r) = (0_K;dr + kr)$ and $(0_K;r)(k;d) = (0_K;rd + kr)$ are valid. Hence, $(0_K;r) \in N_{(K,A)}((K,D)) \cap (\{0_K\} \times rad(A))$ is true. Because of part (ii) of corollary 3 and part (iii) of lemma 2 we deduce that (K, D) is a semisimple K-subalgebra of (K, A). Remark 12 lets us deduce the identity $(0_K;r) \in C_{(K,A)}((K,D)) \cap rad((K,A))$. Hence, for all $d \in D$ $(0_K;d)(0_K;r) = (0_K;r)(0_K;d)$ is true. Based on part (i) of remark 7 we conclude $r \in C_A(D) \cap rad(A)$.

ad(ii): This statements is a consequence of remark 11 and part (i).

ad(iii): The opposite implication is a consequence of part (ii) and of the semsimplicity of D. Let ψ_D be bijective. Based on part (ii) we deduce $C_A(D) \cap rad(A) = \{0_A\}$. $C_A(D) \cap B = C_B(D)$ is valid, and hence $C_B(D) \cap rad(A) = \{0_A\}$ is true, too. Because of $D \subseteq C_B(D)$ we deduce by using Dedekind's identity

$$\begin{aligned} C_B(D) &= \\ B \cap C_B(D) &= \\ (D + rad(A)) \cap C_B(D) &= \\ D + (rad(A) \cap C_B(D)) &= \\ D. \end{aligned}$$

ad(iv): We only have to prove $N_B(D) \subseteq C_B(D)$ in the case of D being commutative. Let D be commutative and $a \in N_B(D)$. Elements $r \in rad(A)$ and $d \in D$ exist such that $a = r + d$ is valid. Because of $d \in N_B(D)$ we deduce $r \in N_B(D) \cap rad(A)$. $C_A(D) \cap B = C_B(D)$ and $N_A(D) \cap B = N_B(D)$ are true, and part (iv) is a consequence of part (i).

ad(v): Within example 1.3.2 the radical complement $D(n, K)$ is commutative. If we use that every element of the centralizer of $D(n, K)$ in $\Delta_{u,n}$ is centralizing the K-basis $\{e_i \mid i \in \underline{n}\}$ von $D(n, K)$, then it is straightforward to prove that $D(n, K)$ is self-centralizing. Based on part (iii) and (iv) we deduce that $D(n, K)$ possesses maximal orbit length. All radical complements are conjugated based on theorem 15 under the action of $(rad(A); *)$. Hence, all radical complements possess the same orbit. By using the last remark within section 1.3.2 part (v) is proven.

ad(vi): The following argumentation is valid:
$\mid D^{(rad(A))} \mid = 1$

$\iff Stab_{rad(A)^\star}(D) = rad(A)$
$\iff_{(i)} C_A(D) \cap rad(A) = rad(A)$
$\iff rad(A) \subseteq C_A(D)$
$\iff rad(A) \subseteq C_B(D)$.

ad(vii): One implication is straightforward to prove. Let T be of minimal orbit length and S be a separable K-subalgebra of A. Based on corollary 4 an element $r \in rad(A)$ exists such that $S \subseteq T^{(r)}$ is valid. By using part (vi) we deduce $rad(A) \subseteq C_A(T^{(r)}) \subseteq C_A(S)$. Hence, $rad(A) \subseteq C_{S+rad(A)}(S)$ is valid. Again, by using part (vi) we have proven statement (vii).⋄

2.5 Invariant radical complements and Taft's theorem

Within this section we transfer Taft's theorem (theorems 11 and 2) to non-unitary associative algebras. At first we need to extend the definition of G-orthogonal elements to non-unitary associative algebras and to extend the action of a group to the adjunction of an unit. The following statements may be proven by the reader within the exercises.

Definition and remark 5 Let K be a field, A a finite-dimensional associative K-algebra and G a finite group such that all elements of G are acting as auto- resp. anti-automorphism on G by $a.g := ag$ resp. $a.g = -ag$ for all $a \in A$ and $g \in G$. Let $a \in A$ and $x \in Q(A)$. The element a is called G-skew (or G-symmetric) if $a.g = a$ resp. $a.g = -a$ for all auto- resp. anti-automorphism $g \in G$. The quasi regular element x is called G-orthogonal if $x.g = x$ resp. $x.g = -x'$ for all auto- resp. anti-automorphism $g \in G$. The collection of all G-orthogonal elements form a subgroup of $Q(A)$ which acts on the G-invariant subalgebras and radical complements. As a consequence the set of all G-orthogonal elements of $rad(A)^\star$ form a subgroup of $rad(A)^\star$ which acts on the G-invariant subalgebras and radical complements.

This definition is compatible with the one presented for unitary algebras because of proposition 3 we know that $1 + x \in E(A)$ and $(1+x)^{-1} = 1 + x'$ for all $x \in Q(A)$ are true. In addition, $1g = 1$ is valid for all $g \in G$.

Let us focus on the adjunction of an unit (K, A). G is also acting on (K, A) by $(k; a).g := (k; a.g)$ for all $a \in A$, $g \in G$ and $k \in K$.

Theorem 17 *(star-invariant radical complements) Let K be a field, A a finite-dimensional associative K-algebra possessing a separable factor algebra by its nilradical and G a finite group such that all elements of G are acting as auto- resp. anti-automorphism on G by $a.g := ag$ resp. $a.g = -ag$ for all $a \in A$ and $g \in G$. The following statements are valid:*

Non-unitary algebras   77

(i) *A* possesses a *G-invariant radical complement*.

(ii) If $char(K) \neq 2$, then the group of *G-orthogonal elements within* $rad(A)^*$ acts transitive on the set of all *G-invariant radical complements of A* by conjugation.

(iii) If $char(K) \neq 2$, then every *G-invariant separable subalgebra T of A* can be conjugated *G-orthogonally based on an elements of* $rad(A)^*$ into a *G-invariant radical complement of A*.

Proof. ad(i)+(iii): We focus on the unitary algebra (K, A). According to corollary 3 and definition and remark 5 the assumptions of theorems 11 and 2 are valid. Thus, (K, A) possesses a G-invariant radical complement X. X is an unital subalgebra based on remark 4. The unital subalgebras of (K, A) are in 1-1 correspondence with the subalgebras of A based on lemma 13 in [56]. Thus, a subalgebra T of A exists such that $X = (K, T)$. X is G-invariant, and hence T is G-invariant, too, by definition and remark 5. Again, by using corollary 3 the subalgebra T is a radical complement of A.

Let S be a G-invariant separable subalgebra of A. (K, S) is a G-invariant separable subalgebra of (K, A) based on corollary 3 and definition and remark 5. We use theorem 2 and deduce that a G-orthogonal element $1 + r$, $r \in rad((K, A))$ exists such that $(K, S)^{1+r} \subseteq (K, T)$ is valid. The unit element of (K, A) is $(1, 0)$, and we use corollary 3 for the existence of an element $x \in rad(A)$ such that $r = (0, x)$ is valid. Based on proposition 3 we know $(1, x)^{-1} = (1 + r)^{-1} = 1 + r'$ and $(0, x)' = (0, x')$. Hence, as proven within theorem 15, the identity $(K, S)^{1+r} = (K, S^{(x)})$ is valid. We deduce $S^{(x)} \subseteq T$.

ad(ii): Part (ii) is a direct consequence of part (iii).◇

2.6 Compatibility properties

Within this section we analyze the compatibility of the theorem of Wedderburn-Malcev with ideals and factor algebras. By compatibility we expect to construct radical complements of ideals and factor algebra by using radical complements of the underlying algebra. For subalgebras we have proven within example 1 that a 4-dimensional associative unitary F-algebra A exists which possesses no radical complement. By using the right regular representation A is isomorphic to a K-subalgebra of $F^{4 \times 4}$, and hence also to one of $F^{4 \times 4} \oplus rad(\Delta_{u,4})$. The latter F-algebra possesses a radical complement but not the isomorphic image of A within it. If the algebra is solvable, then the theorem of Wedderburn-Malcev is also compatible with subalgebras which we will analyze within chapter 3. In addition, within chapter 5 we focus on centers of algebras and analyze the intersection of radical complements with the center. We begin this section by proving elementary properties of the conjugation with quasi regular elements.

Remark 13 *Let A be an associative K-algebra, I a K-ideal and T a K-subalgebra of A. The following statements are valid:*

(i) For all $a \in Q(A)$ the identity $I^{(a)} = I$ is true.

(ii) For all $r \in Q(A)$ the statements $r + I \in Q(A/I)$ and $(r + I)^{(-1)} = r^{(-1)} + I$ are valid.
In particular, if $I \subseteq T$ is valid, then for all $r \in Q(A)$ the identity $(r + I)^{(-1)} * (T/I) * (r + I) = T^{(r)}/I$ is true.

(iii) For all $r \in Q(A)$ the identity $(r+I)^{(-1)}*((T+I)/I)*(r+I) = T^{(r)}+I/I$ is valid.

Proof. ad(i): Let $a \in Q(A)$. By using part (ii) of definition 3 we deduce $I^{(a)} \subseteq I$. Because of remark 9 the set $I^{(a^{(-1)})}$ is a K-ideal of A. By using the inclusion just proven and remark 9 we conclude part (i).

ad(ii): This part can be proven by a straightforward calculation.

ad(iii): This statement is a consequence of part (i), part (ii) and remark 9.◇

Theorem 18 *(compatibility with ideals and factor algebras) Let K be a field, A an associative finite-dimensional K-algebra possessing a separable factor algebra by its nilradical and I an K-ideal of A. The following statements are valid:*

(i) I and A/I are possessing a separable factor algebra by their nilradicals.

(ii) For every radical complement T of A is
$\{(T^{(r)} + I)/I \mid r \in rad(A)\}$ the set of all radical complements of A/I.

(iii) For every radical complement T of A is
$\{T^{(r)} \cap I \mid r \in rad(I)\}$ the set of all radical complements of I.

Proof. ad(i): It is well-known that $rad(I) = rad(A) \cap I$ is valid. Hence, $I/rad(I)$ is $\mathcal{A}$-isomorphic to the K-ideal $(I + rad(A))/rad(A)$ of $A/rad(A)$. Based on part (i) of proposition 2 we deduce the separability of $I/rad(I)$. In addition, it is well-know that $rad(A/I) = (rad(A)+I)/I$ is valid. Hence, $(A/I)/rad(A/I)$ is $\mathcal{A}_1$-isomorphic to $(A/rad(A))/(rad(A) + I/rad(A))$ of the separable K-algebra $A/rad(A)$. Because of part (i) of proposition 2 we deduce (i).

Non-unitary algebras

ad(ii): At first we prove:

(∗) $(T + I)/I$ is a radical complement of A/I.

The K-subalgebra $(T + I)/I$ is $\mathcal{A}_1$-isomorphic to $T/(T \cap I)$ of the separable K-algebra T, and thus it is – based on part (i) of proposition 2 – separable. In particular, $rad(A/I) \cap (T + I)/I = \{0\}$ is valid. Because of $A = rad(A) \oplus_K T$ and $rad(A/I) = (rad(A) + I)/I$ we deduce (∗).

Part (i), (∗) and theorem 15 let us deduce that the set of all radical complements of A/I is exactly the orbit of $(T + I)/I$ under the action of $(rad(A/I); *)$. Because of $rad(A/I) = (rad(A) + I)/I$ and part (iii) of remark 13 statement (ii) is proven.

ad(iii): Based on theorem 13 and part (i) an algebra complement S of $rad(I)$ in I exists. This complement is a separable K-subalgebra of A. By using part (i) of theorem 4 we deduce that an element $r \in rad(A)$ exists such that $S^{(r)} \subseteq T$ is true. $rad(I) = rad(A) \cap I$ is a K-ideal of A. Based on remark 9 and part (i) of remark 13 the set $S^{(r)}$ is an algebra complement of $rad(I)$ in I. In particular, $S^{(r)} \subseteq T \cap I$ is valid. $T \cap I$ is a K-ideal of the separable K-algebra T, and thus by using part (i) of proposition 2 it is a separable K-subalgebra of I. Part (ii) of corollary 4 lets us deduce that $S^{(r)} = T \cap I$ is valid. By this and theorem 15 we conclude that the set of all radical complements of I is exactly the orbit of $T \cap I$ under the action of $(rad(I); *)$. Based on part (i) of remark 13 the theorem is proven.◇

The following diagram illustrates the compatibility of theorem of Wedderburn-Malcev with ideals and factor algebras:

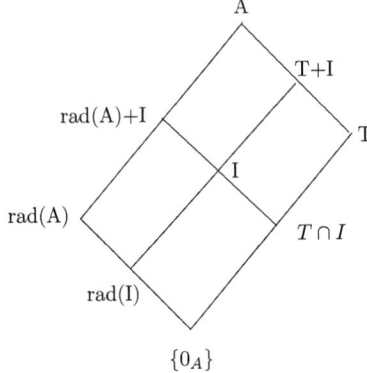

2.7 Top down calculation

Within this section let K be a field and A an associative finite-dimensional algebra possessing a separable factor algebra by its nilradical. Based on theorems 13 and 4 the algebra A possess a radical complement and all radical complements are conjugated under the star group $(rad(A); \star)$.

An algorithm for calculating a radical complement - the case of a zero radical

The first part of this section - the case $rad(A)^2 = 0$ - is based on the proof presented in [8], chapter 6, pages 107-108. We start the calculation with some preliminary remarks.
A radical complement is isomorphic to $A/rad(A)$. If A contains a subalgebra T isomorphic to $A/rad(A)$, then $T \cap rad(A) = 0$ because T is semisimple and $T \cap rad(A)$ is a nilpotent ideal contained in $rad(T)$. Comparing dimensions we derive $A = T \oplus rad(A)$. Thus, A possesses a radical complement if and only if A contains a subalgebra isomorphic to $A/rad(A)$.
Let $\pi : a \mapsto a + rad(A)$ be the natural algebra epimorphism from A onto $A/rad(A)$. Another description of the existence of a radical complement is presented in terms of so-called liftings. A lifting of $A/rad(A)$ is an algebra homomorphism $\epsilon : A/rad(A) \longrightarrow A$ such that $\epsilon \pi = id_{A/rad(A)}$. If A possesses a lifting, then this lifting is injective and A possess a subalgebra isomorphic to $A/rad(A)$ which is a radical complement as already proven. Now let T be a radical complement. T is isomorphic to $A/rad(A)$ as algebra by the function $\beta : t \mapsto t + rad(A)$. It is straightforward to check that $\epsilon := (\beta)^{-1}$ is a lifting of $A/rad(A)$.
As done within the proof of the Wedderburn-Malcev theorem within the appendix the case of $rad(A)^2 = 0$ is essential. The idea is to describe the existence of a radical complement by a solution of a linear equation. We choose a basis $B := \{a_1, \ldots, a_n\}$ of A such that $B_{A\pi} := \{a_1\pi, \ldots, a_m\pi\}$ resp. $B_J := \{a_{m+1}, \ldots, a_n\}$ is basis of $A/rad(A)$ resp. of $rad(A)$. By (α_{ij}^k) we denote the structure constants of A which are elements of K such that for all $i, j \in \underline{n}$ the equation

$$a_i \cdot a_j = \sum_{k=1}^{n} \alpha_{ij}^k \cdot a_k$$

is valid. We conclude

$$a_i\pi \cdot a_j\pi = \sum_{k=1}^{m} \alpha_{ij}^k \cdot a_k\pi.$$

We analyze the conditions for the existence of a lifting ϵ of $A/rad(A)$ into A. A lifting is purely determined by the image of the basis. The condition

Non-unitary algebras

$\epsilon\pi = id$ is equivalent to the fact that for all $i \in \underline{n}$ the equation

$$a_i\pi\epsilon = a_i + \sum_{j=m+1}^{n} x_{ij}a_j$$

is valid for suitable $x_{ij} \in K$. Furthermore, ϵ is an algebra homomorphism. Therefor, for all $i,j \in \underline{m}$ the rule

$$(a_i\pi \cdot a_j\pi)\epsilon = ((a_i\pi)\epsilon) \cdot ((a_j\pi)\epsilon)$$

is true. We evaluate both sides of the equation and use that $rad(A)^2 = 0$ and $rad(A)$ is an ideal of A:

$$(a_i\pi \cdot a_j\pi)\epsilon =$$
$$(\sum_{k=1}^{m} \alpha_{ij}^k \cdot a_k\pi)\epsilon =$$
$$\sum_{k=1}^{m} \alpha_{ij}^k(a_k + \sum_{l=m+1}^{n} x_{kl}a_l) =$$
$$\sum_{l=1}^{m} \alpha_{ij}^l a_l + \sum_{k=1}^{m}\sum_{l=m+1}^{n} \alpha_{ij}^k x_{kl} a_l.$$

$$((a_i\pi)\epsilon) \cdot ((a_j\pi)\epsilon) =$$
$$(a_i + \sum_{k=m+1}^{n} x_{ik}a_k)(a_j + \sum_{k=m+1}^{n} x_{ik}a_k) =$$
$$\sum_{l=1}^{m} \alpha_{ij}^l a_l + \sum_{k,l=m+1}^{n} x_{jk}\alpha_{ik}^l a_l + \sum_{k,l=m+1}^{n} x_{ik}\alpha_{kj}^l a_l.$$

Comparing the coefficients we obtain that a lifting ϵ exists if and only if the following system of linear equation has a solution in K:

$$\sum_{k=1}^{m} \alpha_{ij}^k x_{kl} = \alpha_{ij}^l + \sum_{k=m+1}^{n} \alpha_{ik}^l x_{jk} + \sum_{k=m+1}^{n} \alpha_{kj}^l x_{ik} \text{ for all } i,j \in \underline{n} \text{ and } l \in \underline{n} \setminus \underline{m}.$$

Thus, a lifting and therefor also a radical complement can be calculated based on a system of linear equations over the field K. The linearity of the equation is derivable from the zero radical condition. A solution exists if $A/rad(A)$ is separable based on the Wedderburn-Malcev theorem.◊

An algorithm for calculating a radical complement - the algorithm

We use frequently the previous part in which a calculation of a radical complement is obtained for zero radicals. We start with the factor algebra

$A/rad(A)^2$. The radical of $A/rad(A)^2$ is equal to $rad(A)/rad(A)^2$ which is a zero algebra. The factor algebra $(A/rad(A)^2)/(rad(A)/rad(A)^2)$ is isomorphic to $A/rad(A)$. Thus, we obtain a radical complement C of $A/rad(A)^2$ such that $A/rad(A)^2 = rad(A)/rad(A)^2 \oplus C$ is valid. A subalgebra T_1 of A exists such that $T_1/rad(A)^2 = C$ is true. T_1 is the pre-image under $a \mapsto a + rad(A)^2$. The nilradical of T_1 is equal to $rad(A)^2$ and its factor algebra by the nilradical is isomorphic to $A/rad(A)$. So we can continue with T_1 instead of A. By using the same approach we obtain a subalgebra T_2 of A such that $rad(T_2) = rad(A)^4$ and $T_2/rad(T_2)$ is isomorphic to $A/rad(A)$. This approach is to be done as long as $2^n \geq cl(rad(A))$ is valid. In the n-th step we calculate a subalgebra T_n of A such that $rad(T_n) = rad(A)^{2^n}$ and $T_n/rad(T_n)$ is isomorphic to $A/rad(A)$ are valid. But $rad(A)^{2^n} = 0$ is valid, and thus T_n is a subalgebra of A isomorphic to $A/rad(A)$.◇

Another algorithm

Within the article [13] another algorithm for calculating a radical complement is presented. The algorithm starts with a collection of elements $\{a_1, \ldots, a_k\}$ such that $\{a_1 + rad(A), \ldots, a_k + rad(A)\}$ is a basis of $A/rad(A)$. The authors work in each iteration i with subspaces V_i such that $rad(A)^i = rad(A)^{i+1} \oplus V_i$ is valid. Let (c_{ij}^k) be the structure constants of A. Thus,

$$a_i a_j \equiv \sum_{s=1}^{k} c_{ij}^s a_s \ (mod\, rad(A))$$

is valid for all $i, j \in \underline{k}$. The idea is to transform the elements $a_i, i \in \underline{k}$ to new elements a_i', a_j' such that

$$a_i' a_j' \equiv \sum_{s=1}^{k} c_{ij}^s a_s' \ (mod\, rad(A)^2)$$

is valid for all $i, j \in \underline{k}$, and to do this transformation as long as we reach $rad(A)^{cl(rad(A))} = 0$. The resulting elements form a basis of a subalgebra isomorphic to $A/rad(A)$. Let $b_i^t, i \in \underline{k}$ be the transformed elements after the t-th iteration. Our assumptions is that

$$b_i^t b_j^t \equiv \sum_{s=1}^{k} c_{ij}^s b_s^t \ (mod\, rad(A)^t)$$

is valid for all $i, j \in \underline{k}$. We want to determine elements $\delta_i^t \in V_t$ such that with

$$b_i^{t+1} = b_i^t + \delta_i^t$$

Non-unitary algebras

we have
$$b_i^{t+1}b_j^{t+1} \equiv \sum_{s=1}^{k} c_{ij}^s b_s^{t+1} \pmod{rad(A)^{t+1}}$$

for all $i, j \in \underline{k}$. After expanding this congruence and using the relations
$$\delta_i^t \delta_j^t \equiv 0 \pmod{rad(A)^t}$$

we obtain
$$b_i^t \delta_j^t + \delta_i^t b_j^t \equiv \sum_{s=1}^{k} c_{ij}^s b_s^t - b_i^t b_j^t \pmod{rad(A)^{t+1}}.$$

The Wedderburn-Malcev theorem applied to $A/rad(A))^{t+1}$ implies that there always exist such elements δ_i and conversely a solution of this congruence determine a radical complement in $A/rad(A)^{t+1}$. As mentioned earlier the algorithm stops after reaching the nilpotency class of the radical of A.◇

Calculating another radical complement

Let T be a radical complement. All other radical complements are conjugates of T under the action of $(rad(A); \star)$. Let $x \in rad(A)$. Based on proposition 3 the star inverse of x is the element
$$x' := \sum_{i=1}^{cl(x)} (-1)^x x^i.$$

An upper bound for $cl(x)$ is $cl(rad(A))$ which can be bounded by $dim_K(rad(A)) \leq dim_K(A)$. Now take a basis of B of T. The set $B^{(x)}$ is a basis for the radical complement $T^{(x)}$. For each element $b \in B$ we calculate based on proposition 3 the identity
$$b^{(x)} = x' \star b \star x = b + x'b + bx + x'bx.◇$$

Calculating a decomposition for an element and the transfer rule

Let T be a radical complement, $a \in A$ and $x \in rad(A)$. One decomposition of a related to $rad(A)$ and T can be obtained by using a basis for $rad(A)$ and for T and solving a system of linear equation. The solution is a decomposition $a = r + t \in rad(A) \oplus T$. The same can be done for $T^{(r)}$ instead of T. But the following transfer rule can be used to transfer the decomposition $a = r + t$ to $rad(A) \oplus T^{(r)}$. We have already used the rule
$$t^{(x)} = x' \star t \star x = t + x't + tx + x'tx.$$

Thus,
$$t^{(x)} - t = x't + tx + x'tx \in rad(A)$$
is true because $x \in rad(A)$ and $rad(A)$ is an ideal of A. We derive the transfer rule
$$a = r + t = (r + t^{(x)} - t) + t^{(r)} \in rad(A) \oplus T^{(r)}.\diamond$$

2.8 Open-ended questions and exercises

Open-ended questions 1 *(i) Determine the action of the star group on the set of separable subalgebras.*

(ii) Eiichi Abe proves in [1] the theorem of Wedderburn-Malcev for coalgebras possessing a separable coradical. Is it possible to prove this result without assuming that the coalgebra possesses a counit?

(iii) For which algebras every semisimple subalgebra is separable? We have proven this for solvable algebras possessing a separable factor algebra by its nilradical.

(iv) Under the conditions of the Wedderburn-Malcev theorem analyze whether maximal semisimple subalgebras are separable.

Excercise 111 *Apply remark 13 to unitary algebras.*

Excercise 112 *Generalize theorem 18 to G-invariant ideals within the context of Taft's theorem.*

Excercise 113 *Analyze whether/how Taft's theorem can be enhanced to the following algebras:*

(i) A direct product of associative algebras.

(ii) A tensor product of associative algebras.

(iii) A full matrix algebra over an associative algebra.

(iv) A G-invariant subalgebra, left and right-ideal of a G-invariant associative algebra.

(v) A G-invariant subalgebra of an solvable associative algebra.

Excercise 114 *Let K be a field, A an associative finite-dimensional K-algebra, Z a central separable subalgebra and H a semisimple subalgebra of A. Prove that the generated subalgebra of H and Z is semisimple. Use*

Non-unitary algebras

that the tensor product of a separable and semisimple algebra is semisimple. Deduce that the every central and separable subalgebra is contained in the intersection of all maximal semisimple subalgebras of A. What is the consequence for the radical complement of the center of solvable associative algebra if $A/rad(A)$ is separable?

Excercise 115 Let K be a field and A an associative finite-dimensional K-algebra. The following statements are valid:

(i) Let H be a semisimple subalgebra of A. Prove $dim_K(H) \leq dim_K(A/rad(A))$.

(ii) Prove that the maximal dimension of all separable subalgebras of A is not greater than the maximal dimension of all semisimple subalgebras of A. The latter dimension is not greater than $dim_K(A/rad(A))$.

(iii) Let $A/rad(A)$ be separable. All dimensions in item (ii) are identical.

Excercise 116 Within the context of corollary 4 prove that the radical complements are maximal semisimple subalgebras related to $\subseteq$ and related to dimension.

Excercise 117 Within theorem 16 clarify if the following statement is correct: A radical complement possesses maximal orbit length if and only if every separable subalgebra possesses maximal orbit length. (Hint: center of the algebra)

Excercise 118 Enhance theorem 16 and prove that the orbit length of the radical complements are maximal among all separable and semisimple subalgebras.

Excercise 119 Analyze the article [11] and prove all statements.

Excercise 120 Analyze the proof presented within [8] for the existence part of the Wedderburn-Malcev theorem.

Excercise 121 Analyze the proof presented within [8] for the conjugacy part of the Wedderburn-Malcev theorem.

Excercise 122 Apply the whole section 2.7 to the algebra of lower triangular matrices in three dimensions. Use the subalgebra of diagonal matrices and two other radical complements for the calculations.

Excercise 123 Is it possible to transfer every algebraic statement by using the adjunction of an unit from unitary to non-unitary associative algebras? (Hint: [55], [56])

Excercise 124 Prove all statements within definition and remarks 5.

Excercise 125 *Generalize exercise 40 to non-unitary algebras using the star composition.*

Excercise 126 *Let A be a finite-dimensional associative K-algebra possessing a separable factor algebra by its nilradical. For each radical complement T of (K, A) focus on $T \cap A\varphi$. Analyze whether this map is injective, surjective or bijective. Determine the cardinality of the set of radical complements of A and of (K, A) and compare them. What is the result if A is unitary?*

Excercise 127 *Is a Lie algebra unitary? Is a Jordan algebra unitary? If it is necessary, then do a research in the literature for the corresponding definitions of Lie and Jordan algebras.*

Excercise 128 *Is the adjunction of an unit with a Lie resp. Jordan algebra again a Lie resp. Jordan algebra?*

Excercise 129 *Let A be an associative K-algebra. For every idempotent element e of A prove that eAe is an associative unitary K-algebra with unit element e. eAe is an unitary K-subalgebra. On what terms is eAe unital? Do additional unitary K-subalgebras of A exist? Do additional unitary one-sided K-ideals of A exist except eA and Ae?*

Excercise 130 *Let A be an unitary K-algebra. $K \cdot 1_A$ is an unital K-subalgebra of A. Is it unital, too?*

Excercise 131 *Prove theorem 12 by using the cited literature.*

Excercise 132 *Analyze the following K-algebras whether unitary and proper unital and unitary K-subalgebras exist:*

  (i) $K^{n \times n}$, K a field, $n \in \mathbb{N}$

  (ii) $A^{n \times n}$, A an associative unitary K-algebra, $n \in \mathbb{N}$

  (iii) the $\mathbb{Q}$-algebra of lower triangular matrices over $\mathbb{Q}$

  (iv) the $\mathbb{R}$-algebra of upper triangular matrices over $\mathbb{R}$

  (v) $\mathbb{C}^n$, $n \in \mathbb{N}$

  (vi) the set of strict upper triangular matrices over $\mathbb{Z}$

  (vii) the set of strict lower triangular matrices over $GF(2)$.

Excercise 133 *Is a zero-algebra unitary or separable?*

Excercise 134 *True or false: Ideals and factor algebras of separable algebras are separable.*

Excercise 135 A K-algebra is called local if and only if the factor algebra by its nilradical is a K-division algebra. Let N be a finite-dimensional associative nilpotent K-algebra based on a field K. Prove that N is not unitary and that the adjunction of unit (N, K) is local. Describe the corresponding K-division algebra of the factor algebra by its nilradical. Determine the nilradical of (K, N). Is the factor algebra by its nilradical a separable K-algebra? Try to determine the cardinality of all radical complements of (K, N). Is every associative unitary local algebra isomorphic to an adjunction of an unit if the radical factor algebra is one-dimensional?

Excercise 136 Solve exercise 135 again for a zero-algebra N.

Excercise 137 Let K be a field, $char(K) = p > 0$ and G be a finite p-group. By using a theorem of Wallace prove that the augmentation ideal of KG is nilpotent. ($Aug(KG)$ is K-generated by the set $\{g-1_G \mid g \in G\}$.) Determine the dimension of $Aug(KG)$. Prove that $K1_G$ is a radical complement and that $Aug(KG)$ is the nilradical. Analyze whether a connection exists between KG and the K-algebra $(Aug(KG), K)$ (tip: exercise 135).

Excercise 138 Let A be a finite-dimensional associative K-algebra and L resp. R a separable left resp. right ideal of A. Prove that $L + R$ is a K-subalgebra of A. Is the intersection $L \cap R$ separable? Is the intersection $L \cap R$ an ideal of $L + R$? Is $L + R$ separable?

Excercise 139 Formulate the identity of Dedekind for algebras and prove it.

Excercise 140 We focus on the algebra $\Delta_{u,3}$ over the field $GF(p)$. Determine and count all radical complements. Find a conjecture for $\Delta_{u,n}$ as algebra over an arbitrary finite field.

Excercise 141 Let A be a finite-dimensional associative K-algebra possessing a separable factor algebra by its nilradical and let T be a radical complement. For every element $r \leq cl(rad(A))$ let us focus on the ideal $rad(A)^r$. Determine the nilradical and – by using T – a radical complement of $A/rad(A)^r$. In addition, determine the class of nilpotency for the nilradical of $A/rad(A)^r$. ($cl(A)$ is the class of nilpotency for an associative K-algebra A. For every element $a \in A$ we call $cl(a)$ the class of nilpotency of a.)

Excercise 142 Solve exercise 141 again for the algebra $\Delta_{o,4}$ over an arbitrary finite field.

Excercise 143 Let A be an associative K-algebra. For every element $a \in A$ and $n \in \mathbb{N}$ calculate the n-th power of a with respect to $\star$. In addition, do this calculation for $char(K) = p$ and $n = p^r$, $r \in \mathbb{N}$.

Excercise 144 Let A be an associative K-algebra and $a, b, c \in A$. Calculate $a \star b \star c$. Find a conjecture for finite many elements.

Excercise 145 Let N be a nilpotent associative K-algebra. True or false: (K, N) and $K \times N$ are isomorphic.

Excercise 146 Let A be an associative K-algebra. Describe all units of (K, A) by using the star composition $\star$. Does a connection between $Q(A)$, $Q(K, A)$ and $E(K, A)$ exist?

Excercise 147 Prove the statements (ii) and (iii) of lemma 2.

Excercise 148 Let $N := rad(\Delta_{u,3})$, $A := (K, N)$ and $I := rad(N)^2$. To which algebra is $(K, N)/(0, I)$ isomorphic?

Excercise 149 Let A be an associative K-algebra with $rad(A) = 0$. For an element e of A prove the equivalence of the following statements:

(i) e is a left-sided unit of A.

(ii) e is a right-sided unit of A.

(iii) e is an unit of A.

Is this equivalence true without assuming $rad(A) = 0$? (Tip: prove of theorem 1, use A^-))

Excercise 150 Determine all units, left-sided and right-sided units of $\Delta_{u,2}$ and $\Delta_{o,3}$.

Excercise 151 Let A be an associative K-algebra. What is the meaning of an idempotent element e of A for the K-subalgebras eA and Ae?

Excercise 152 Do idempotent elements e of $\Delta_{u,2}$ exist such that e is an unit of eA? Is this set of idempotents a structure?

Excercise 153 Do idempotent elements e of $\Delta_{u,2}$ exist such that e is an unit of Ae? Is this set of idempotents a structure?

Excercise 154 Let A be an associative finite-dimensional K-algebra and I an ideal of A. Determine the nilradical of I and of A/I.

Excercise 155 Prove lemma 3 in details.

Excercise 156 Do a research in the literature and prove lemma 5.

Excercise 157 Do a research in the literature (tip: G. Karpilovsky) for the statement that the group algebra KG – based on a field K and on a finite group G – possesses always a separable factor algebra $KG/rad(KG)$.

Non-unitary algebras   89

Excercise 158 *Let D and E central finite-dimensional associative K-division algebras possessing coprime dimensions. Prove that their tensor product is again a central finite-dimensional associative K-division algebra (tip: see [29] or [35]). What is the dimension of the tensor product?*

Excercise 159 *For theorem 14 analyze how to bound the class of nilpotency of the nilradical of the mentioned tensor product by the corresponding factors.*

Excercise 160 *Solve example 4 for a non-unitary associative algebra.*

Excercise 161 *On what terms is the shift by an element a homomorphism on an Abelian group?*

Excercise 162 *For the following algebras analyze whether the statement $A \otimes A \cong A \oplus A$ is true or false: $\mathbb{Q}[i]$, K^4, the algebra within example 5, the first column-space of $K^{2\times 2}$, the first row-space of $K^{3\times 3}$, the center of $\Delta_{u,6}$, a radical complement of $\Delta_{o,8}$, a subfield of a finite field regarded as algebra over its prime field.*

Excercise 163 *Prove proposition 3 in details (if needed, then by using the cited literature).*

Excercise 164 *Prove remark 8 in details.*

Excercise 165 *Prove within remark 9 the statement that $Z(A) \cap Q(A)$ is exactly the kernel.*

Excercise 166 *Formulate and prove an unitary version of corollary 4.*

Excercise 167 *Prove remark 13 in details.*

Excercise 168 *Let A be an associative finite-dimensional K-algebra and T a K-subalgebra of A each possessing a separable factor algebra by their nilradical. Prove that a radical complement X of A exists such that $X \cap T$ is a radical complement of T. Is this statement true in general for all radical complements of A? Is it possible to enhance each radical complement of T to one of A?*

Excercise 169 *Let A be an associative finite-dimensional K-algebra and T a K-subalgebra of A each possessing a separable factor algebra by their nilradical. Let us assume $rad(A) = rad(T)$. Prove for every radical complement X of A that $X \cap T$ is a radical complement of T.*

Excercise 170 *Calculate an example for $\Delta_{u,n}$ within exercise 169 in details.*

Excercise 171 *Let us focus on example 5 based on a finite field K. Count the number of radical complements. What is the cardinality of a radical complement? Determine the orbits and their length of the set of separable subalgebras under the action of the star group of the nilradical. What is the effect of the field K on the solution of this exercise?*

Excercise 172 *For the following algebras determine the nilradical, one radical complement, the dimensions of these structures and try to describe all radical complements:*

(i) $\Delta_{u,3}$ over $\mathbb{Q}$

(ii) $\Delta_{u,3} \times s\Delta_{u,3}$ over $GF(7)$

(iii) $\Delta_{o,4} \times \Delta_{o,4}$ over $\mathbb{C}$

(iv) $\Delta_{u,5} \otimes \Delta_{o,5}$ over $\mathbb{Q}(i)$

(v) the column-space related to the first column over $GF(11)^{7\times 7}$

(vi) the row-space related to the first row over $GF(13)^{8\times 8}$

(vii) the direct product related to the first column- and row-space over $\mathbb{R}^{6\times 6}$.

Which of these algebras are unitary?

Excercise 173 *(eAe) Let K be a field, A an associative finite-dimensional unitary K-algebra possessing a separable factor algebra by its nilradical, T a separable radical complement of $rad(A)$ in A and e an idempotent element of A. Prove the following statements:*

(i) e is diagonalizable and hence separable.

(ii) The subalgebra K-generated by the set $\{1, e\}$ is separable, and thus is contained in a conjugate of T based on an element $1 + r$, $r \in rad(A)$.

(iii) Do a research in the literature and prove that $e\, rad(A)\, e$ is the nilradical of eAe and prove this statement afterwards.

(iv) Use $A = rad(A) \oplus T^{1+r}$ to deduce $eAe = e\, rad(A)\, e \oplus eT^{1+r}e$. In addition prove that $eT^{1+r}e$ is a separable radical complement of eAe.

(Tip for item (iv): Use theorem 1 and part (iii).)

Excercise 174 *(zero-extension) Let A be a K-algebra based on the multiplication $\cdot$. On the set $B := A \times A$ we define a new multiplication $\odot$ by $(a, x) \odot (b, y) := (ab, ay + xb)$. Prove and analyze the following statements:*

(i) $(B; \odot)$ is a K-algebra.

Non-unitary algebras

(ii) $(A;\cdot)$ is associative if and only if $(B;\odot)$ is associative.

(iii) $(A;\cdot)$ is commutative if and only if $(B;\odot)$ is commutative.

(iv) If $(A;\cdot)$ is unitary, then $(B;\odot)$ is unitary, too, and $(1_A;0)$ is an unit.

(v) Is the opposite inclusion also true within the previous item?

(vi) The center of $(B;\odot)$ is $Z(A) \times Z(A)$. Is $\odot$ equal to $\cdot$ on the center?

(vii) $0 \times A$ is a nilpotent ideal of B for which all products are zero (a so-called zero-ideal). Is $\odot$ equal to $\cdot$ on $0 \times A$?

(viii) $(B/(0 \times A);\odot)$ is isomorphic to $(A;\cdot)$.

(ix) Let A be an associative unitary finite-dimensional separable K-algebra. B is an unitary finite-dimensional associative K-algebra possessing the nilradical $0 \times A$ and a separable factor algebra by the nilradical which is isomorphic to $(A;\cdot)$. $(A \times 0;\odot)$ is a radical complement. Describe $\odot$ on this radical complement.

(x) What would be a conjecture within the previous item if an algebra of the form $A = rad(A) \oplus T$ is used possessing a separable radical complement T? Is it possible to prove this conjecture?

(xi) Is $(B;\odot)$ isomorphic to $(A \times A;\cdot)$?

(xii) Describe the multiplication of the inverse algebra of B. In what way is it similar to $\odot$?

Excercise 175 Calculate all statements within the exercises 174 and 173 by using the algebra $\Delta_{u,3}$ in the following way: for the zero-extension use the algebra of diagonal matrices, within the second exercise find an idempotent element different from 1 and 0. What are the idempotent elements of $\Delta_{u,3}$?

Excercise 176 Analyze the consequence within exercise 174 if the multiplication is replaced by $(a,x) \odot (b,y) := (ay + xb, ab)$.

Excercise 177 Solve again exercise 100 by using statements about nilpotency and quasi regularity.

Excercise 178 Let a be an element of an associative K-algebra A. True or false: a is nilpotent if and only if $K[a]$ is nilpotent. If a is nilpotent, then $cl(a) = cl(K[a])$ is valid.

Excercise 179 Prove that ideals and factor algebras of semisimple associative algebras are semisimple but right and left ideals need not to be semisimple.

Excercise 180 *Prove theorem 1 again by using right-sided units instead of left-sided units.*

Excercise 181 *Let A be an associative algebra and e a left-sided unit of A. Prove that e is an idempotent of A and that $\{x \mid x \in A, xe = 0\}$ is a zero right ideal of A. Is $\{x \mid x \in A, ex = 0\}$ also a substructure of A?*

Excercise 182 *Let A be an associative algebra and e a right-sided unit of A. Prove that e is an idempotent of A and that $\{x \mid x \in A, ex = 0\}$ is a zero left ideal of A. Is $\{x \mid x \in A, xe = 0\}$ also a substructure of A?*

Excercise 183 *True or false: Left ideals of separable algebras are separable.*

Excercise 184 *True or false: Right ideals of separable algebras are separable.*

Excercise 185 *Let A be an associative K-algebra and I a K-ideal of A. For all $n \in \mathbb{N}$ prove that $(A/I)^{n \times n}$ is isomorphic to $A^{n \times n}/I^{n \times n}$.*

Excercise 186 *Let A, B be an finite-dimensional associative K-algebras possessing a separable radical factor algebra. Bound the nilpotency class of $rad(A \otimes B)$. Apply the result to the tensor product $\Delta_{u,n} \otimes \Delta_{o,n}$ for an arbitrary $n \in \mathbb{N}$.*

Excercise 187 *Study the article [2]. Analyze which version of the theorem of Wedderburn-Malcev is used within this article. If the result of theorem 6 would only be valid for unital algebras, then transfer the result to non-unitary algebras.*

Excercise 188 *Study the article [2]. Transfer all results to right ideals by executing the proofs again.*

Excercise 189 *Study the article [2]. Transfer all results to right ideals by using the opposite algebra.*

Excercise 190 *Study the article [2]. In view of theorem 6 find an algebra A, a left ideal I and a radical complement S of A such that $A = rad(A) \oplus S$ and $I = (I \cap rad(A)) \oplus (I \cap S)$ are valid but $I \cap rad(A)$ is not the nilradical of I and $I \cap S$ is not a radical complement of I. (Hint: row resp. line spaces in matrix algebras)*

Excercise 191 *Study the article [2]. In view of theorem 6 find an algebra A, a right ideal I and a radical complement S of A such that $A = rad(A) \oplus S$ and $I = (I \cap rad(A)) \oplus (I \cap S)$ are valid but $I \cap rad(A)$ is not the nilradical of I and $I \cap S$ is not a radical complement of I. (Hint: row resp. line spaces in matrix algebras)*

Chapter 3

Solvable algebras

3.1 Basic properties

We begin this section by defining solvable associative algebras as presented in [3]. After proving a first result (theorem 19) we will see that this definition is related to the definition of solvability for groups and Lie algebras.

Definition and remark 6 *(solvable associative algebra)* Let K be a field. An associative K-algebra A is called solvable (or soluble) if $A/rad(A)$ is commutative. In particular, every associative commutative, every associative nilpotent and the algebras of upper and lower triangular matrices over a field are solvable (because their radical factor algebra is isomorphic to the subalgebra of diagonal matrices). Other prominent examples of solvable associative algebras are the Solomon algebras and the Solomon-Tits algebras which are analyzed e.g. within [3] and [52]. Both play an important role within the representation theory of the symmetric groups.⋄

Remark 14 *Let A be an associative right artian semisimple K-algebra and I an ideal of A. The following statements are valid:*

(i) $Z(I) = Z(A) \cap I$

(ii) A is commutative if and only if I and A/I are commutative.

(iii) If K is a field and A is finite-dimensional, then $Z(A/I) = (Z(A)+I)/I$ is valid.

Proof. ad(i): A is semisimple, and thus the set of minimal K-ideals of A is a finite direct decomposition of A. Some of these ideals are decomposing the K-ideal I. Hence, part (i) is proven.

ad(ii): Because of the semisimplicity of A the K-ideal I possesses an ideal complement in A which is $\mathcal{A}$-isomorphic to A/I. This proves part (ii).

ad(iii): Because of the semisimplicity of A the K-ideal I possesses an ideal complement J in A. The statement $A/I \cong_A J$ is true, and thus $Z(A/I) \cong_A Z(J)$ is valid. In addition, $(Z(A) + I)/I \subseteq Z(A/I)$ is straightforward to prove. By using the statement

$$\begin{aligned}(Z(A)+I)/I &= \\ ((Z(I) \oplus_K Z(J))+I)/I &= \\ (Z(J) \oplus_K I)/I &\cong_A \\ Z(J) \end{aligned}$$

and the finite dimension of A the proof is finished. $\diamond$

Definition 6 *(powers of associative algebras)* Let A be an associative K-algebra and $n \in \mathbb{N}$. By $A^{<n>}$ we denote the K-linear span of all n-fold associative products of A. This span is called the n-th power of A. A is nilpotent if and only if one n-th power is the zero-space. The minimal n such that $A^{<n>}$ is zero is called the class of nilpotency of A. It is denoted by $cl(A)$. $\diamond$

Theorem 19 *(characterization of solvability)* Let A be a associative finite-dimensional K-algebra. The following statements are equivalent:

(i) A is solvable.

(ii) A K-ideal I of A exists such that I is nilpotent and A/I is commutative.

(iii) A K-ideal I of A exists such that I is solvable and A/I is commutative.

(iv) A K-ideal I of A exists such that I and A/I are solvable.

(v) For every K-ideal I of A the algebras I and A/I solvable.

(vi) An element $n \in \mathbb{N}$ and K-ideals I_i ($0 \leq i \leq n$) of A exist such that $\{0_A\} = I_n \leq I_{n-1} \leq \cdots \leq I_0 = A$ is valid and for all $i \in \underline{n}$ the factor algebra I_i/I_{i-1} is commutative.

(vii) An element $n \in \mathbb{N}$ and K-subalgebras I_i ($0 \leq i \leq n$) exists such that $\{0_A\} = I_n \leq I_{n-1} \leq \cdots \leq I_0 = A$ is valid and for all $i \in \underline{n}$ the subalgebra I_{i-1} is a K-ideal of I_i possessing a commutative factor algebra I_i/I_{i-1}.

Proof. The implication from (i) to (ii) is true by using $I := rad(A)$. Nilpotent and commutative associative algebras are solvable, and thus the implications from (ii) to (iii) and from (iii) to (iv) are true. The implication from (vi) to (vii) is straightforward to prove. In addition, the implication from

(v) to (i) is valid by applying part (v) to the K-ideal A.

Now we prove the implication from (i) to (v). Let I be an ideal of A. A is solvable, and by definition $A/rad(A)$ is commutative. It is well-known that $rad(I) = rad(A) \cap I$ and $rad(A/I) = (rad(A)+I)/I$ are valid. By using well-known isomorphism theorems we deduce that $I/rad(I)$ is $\mathcal{A}$-isomorphic to a subalgebra of $A/rad(A)$. Therefor, the algebra $I/rad(I)$ itself is commutative. In addition, $(A/I)/((rad(A) + I)/I)$ is $\mathcal{A}$-isomorphic to the algebra $A/(rad(A) + I)$. $A/rad(A)$ is commutative, and hence $A/(rad(A) + I)$ is commutative, too. Thus, we have proven part (v).

Now we prove the implication from (iv) to (i). Let I be a K-ideal of A such that I and A/I are solvable. We use $rad(I) = rad(A) \cap I$, the solvability of I and well-known algebra isomorphism theorems to prove that the ideal $(rad(A) + I)/rad(A)$ of $A/rad(A)$ is commutative. In addition, $(A/rad(A))/((rad(A) + I)/rad(A))$ is $\mathcal{A}$-isomorphic to the algebra $B := (A/I)/((rad(A) + I)/I)$. By using $rad(A/I) = (rad(A) + I)/I$ and the solvability of A/I we deduce the commutativity of B. Finally, part (ii) of remark 14 is used to prove the solvability of A.

We prove now the implication from (i) to (vi). Let $c := cl(rad(A))$ and $m \in \mathbb{N}$ minimal such that $c \leq 2^m$ is valid. For all $s \in \underline{m}$ the algebra $rad(A)^{<2^s>}$ is an ideal of A. These ideals of A satisfy the statement (vi).

We finish the proof by deducing the implication (vii) to (i). Let part (vii) be valid, $n \in \mathbb{N}$ and I_i ($0 \leq i \leq n$) be K-ideals of A satisfying the statement (vii). A standard induction element lets us deduce that I_1 is solvable. A/I_1 is solvable, too, and thus – using the already proven implication (iv) to (i) – statement (i) is proven.⋄

Corollary 3 *Let A, B, C be a finite-dimensional associative solvable K-algebra. The following statements are valid:*

(i) Every K-subalgebra of A is solvable.

(ii) $B \oplus C$ is solvable if and only if B and C are solvable.

Proof. ad(i): Let T be a K-subalgebra of the solvable K-algebra A. Based on part (vii) of theorem 19 an element $n \in \mathbb{N}$ and K-ideals I_i ($0 \leq i \leq n$) of A exist such that $\{0_A\} = I_n \leq I_{n-1} \leq \cdots \leq I_0 = A$ is valid and for all $i \in \underline{n}$ the algebra I_i/I_{i-1} is commutative. For all $0 \leq i \leq n$ the algebra $I_i \cap T$ is a K-ideal of T, and for all $1 \leq i \leq n$ the algebra $(I_i \cap T)/(I_{i-1} \cap T)$ is commutative. By using part (vii) of theorem 19 the statement (iii) is proven.

ad(ii): The implication $\Longrightarrow$ is a consequence of part (v) of theorem 19.

By using part (iv) of theorem 19 the opposite implication is true.◊

The next corollary let two significant subalgebras arise within every associative algebra. Both subalgebras are describing the deviation of an associative algebra for being solvable.

Corollary 4 *Let A be an associative finite-dimensional K-algebra and I, J two K-ideals of A. The following statements are valid:*

(i) *If I and J are solvable, then $I + J$ is solvable.*

(ii) *If A/I and A/J are solvable, then $A/(I \cap J)$ is solvable.*

(iii) *A possesses a unique maximal solvable K-ideal $AUF(A)$. $rad(AUF(A)) = rad(A)$ and $AUF(A/AUF(A)) = 0$ are valid. The commutative minimal K-ideals of $A/rad(A)$ are decomposing $AUF(A)/rad(A)$ directly. We call $AUF(A)$ the solvable radical of A.*

(iv) *A possesses a smallest K-ideal $auf(A)$ such that $A/auf(A)$ is solvable. We call $auf(A)$ the solvable residuum of A.*

Proof. ad(i): I is a solvable K-ideal of $I + J$. By using $(I + J)/I \cong_A J/(I \cap J)$ and statement (v) of theorem 19 the factor algebra $(I + J)/I$ is solvable. Part (iv) of theorem 19 finishes the proof of part (i).

ad(ii): A/J is solvable, and thus the K-ideal $(I + J)/J$ of A/J is solvable based on part (v) of theorem 19. Hence, the algebra $I/(I \cap J)$ – which is isomorphic to $(I+J)/J$ – is solvable, too. By using the solvability of A/I also the algebra $(A/(I \cap J))/(I/(I \cap J))$ is solvable. Part (iv) of theorem 19 finishes the proof of statement (ii).

ad(iii): By using part (i) the algebra A possesses a unique maximal solvable K-ideal $AUF(A)$.
$rad(A)$ is a nilpotent K-ideal of A, and thus it is contained in $AUF(A)$. $AUF(A)/rad(A)$ is an K-ideal of the semisimple K-algebra $A/rad(A)$. Therefor, $AUF(A)/rad(A)$ is semisimple. We conclude $rad(A) = rad(AUF(A))$. Let I be a K-ideal of A containing $AUF(A)$ such that $I/AUF(A)$ is solvable. $AUF(A)$ is solvable, and thus – by using part (iv) of theorem 19 – I is solvable, too.
$AUF(A)/rad(A)$ is a K-ideal of $A/rad(A)$, and thus some of the minimal ideals of $A/rad(A)$ are decomposing $AUF(A)/rad(A)$. Because of the solvability of $AUF(A)$ and the condition $rad(A) = rad(AUF(A))$ these minimal K-ideals are commutative.
Now let I be the K-ideal of A such that the commutative minimal K-ideals of $A/rad(A)$ are decomposing $I/rad(A)$. We have already proven the statement $AUF(A) \subseteq I$. Vice versa, I is solvable (because $rad(A)$ and $I/rad(A)$

Solvable algebras 97

are solvable, see theorem 19). We conclude $I = AUF(A)$.

ad(iv): Part (iv) is deductable from part (ii).⋄

Our next topic is to define and to analyze the class solvability for an associative solvable algebra which is related to the chain of derived subalgebras. The structural importance is presented and discussed within theorem 20.

Definition and remark 7 *(chain of derived subalgebras, class of solvability)* Let A be an associative K-algebra and S, T subsets of A. By $\langle T \rangle_{\trianglelefteq A}$ we denote the smallest K-ideal of A containing T. A is – with respect to the composition $a \circ b := ab - ba$ – a Lie algebra called the associated Lie algebra of A and is denoted by A°. Furthermore, we define $S \circ T := \langle s \circ t \mid (s,t) \in S \times T \rangle_K$. The chain of derived subalgebras of A is defined by $A^{(0)} := A$, $A' := A^{(1)} := \langle A \circ A \rangle_{\trianglelefteq A}$ and $A^{(n)} := (A^{(n-1)})'$ for all $n \in \mathbb{N}_{\geq 2}$. For all $n \in \mathbb{N}$ the subalgebra $A^{(n)}$ is the smallest ideal of $A^{(n-1)}$ possessing a commutative factor algebra $A^{(n-1)}/A^{(n)}$. We call $A^{(n)}$ the n-th derived subalgebra of A and $A^{(1)} = A'$ the derived subalgebra or also derivation of A.⋄

The structural importance of the chain of derived subalgebras is presented within the next theorem: This chain is the shortest finite chain from zero to the whole algebra possessing successive commutative factor algebras a long this chain.

Theorem 20 *(class of solvability)* Let A be a finite-dimensional associative K-algebra. The following statements are valid:

(i) A is solvable if and only if an element $n \in \mathbb{N}$ exists such that $A^{(n)} = \{0_A\}$ is valid. If A is solvable, then the minimal $n \in \mathbb{N}$ such that $A^{(n)} = \{0_A\}$ is valid is called the class of solvability or solvable class of A denoted by $st(A)$.

(ii) Let A be solvable. If $n \in \mathbb{N}$ and I_i $(0 \leq i \leq n)$ are subalgebras of A such that $\{0_A\} = I_n \leq I_{n-1} \leq \cdots \leq I_0 = A$ is valid and for all $i \in \underline{n}$ the set I_{i-1} is an ideal of I_i possessing a commutative factor algebra I_i/I_{i-1}, then $st(A) \leq n$ is true.

Proof. One implication in (i) is deductable by using theorem 19. Let A be solvable. By definition an element $n \in \mathbb{N}$ and subalgebras I_i $(0 \leq i \leq n)$ of A exist such that $\{0_A\} = I_n \leq I_{n-1} \leq \cdots \leq I_0 = A$ is valid and for all $i \in \underline{n}$ the algebra I_{i-1} is a K-ideal of I_i possessing a commutative factor algebra I_i/I_{i-1}. We prove $A^{(n)} = \{0_A\}$ from which we deduce one part of (i) and statement (ii). For this we use an induction argument to prove that for all $r \in \mathbb{N}$ the condition $A^{(r)} \subseteq I_r \cap A^{(r-1)}$ is true. I_1 is a K-ideal possessing a commutative factor algebra A/I_1. Thus, $A^{(1)} \subseteq I_1 \subseteq I_1 \cap A$ is valid. Let $r \in \mathbb{N}$ satisfying the condition $A^{(r)} \subseteq I_r \cap A^{(r-1)}$. I_{r+1} is a K-ideal of I_r

possessing a commutative factor algebra I_r/I_{r+1}. Hence, $I_{r+1} \cap A^{(r)}$ is a K-ideal of $A^{(r)}$. By using well-known isomorphism theorems we deduce that $A^{(r)}/(I_{r+1} \cap A^{(r)})$ is $\mathcal{A}$-isomorphic to a subalgebra of I_r/I_{r+1}. Hence, it is commutative, too. Thus, we have proven $A^{(r+1)} \subseteq I_{r+1} \cap A^{(r)}$.⋄

3.2 Connections to the associated Lie algebra

3.2.1 A Lie characterization

In modern algebra connecting structure with each other has a long tradition starting with Galois theory. By doing so, statements can be transferred between both structures. This section is based on this concept.
Within [3] solvable associative algebras are connected to solvable groups in terms of their group of units. The argumentation within [3] needs to be enhanced: this is done within the appendix of this work, and in addition, we will generalize the theorem to non-unitary algebras.

Within this section we will characterize solvable associative algebras with their associated Lie algebras. The next proposition shows that the associated Lie algebra of an associative solvable algebra is solvable, too.

Proposition 4 *If A is an associative finite-dimensional solvable K-algebra, then A° is solvable.*

Proof. A is solvable, and thus $A^\circ \circ A^\circ \subseteq rad(A)$ is valid. The nilpotency of $rad(A)$ lets us deduce the nilpotency of $rad(A)^\circ$ (This statement may be proven by the reader as an exercise.). $rad(A)^\circ$ is a K-subalgebra of A°, and thus $A^\circ \circ A^\circ$ is nilpotent. We conclude that $A^\circ \circ A^\circ$ is solvable. Therefor, A° is solvable.⋄

For proving the opposite implication we use a well-known method within the theory of associative algebras: the statement is proven for central division algebras at first, afterwards for arbitrary division algebras, then for simple and semisimple algebras and finally for arbitrary algebras. For executing this method the next remark and lemma are needed. For a Lie algebra L resp. for a group G let $(L^{(n)})_{n \in \mathbb{N}}$ resp. $(G^{(n)})_{n \in \mathbb{N}}$ the series of derived subalgebras of L resp. derived subgroups of G.

Remark 15 *Let K be a field and A, B associative unitary K-algebras. The following statements are valid:*

(i) For all $a, c \in A$ and $b, d \in B$ the identity

$$(a \otimes b) \circ (c \otimes d) = (a \circ c) \otimes bd + ca \otimes (b \circ d)$$

is valid.

(ii) If A° is solvable and B° Abelian, then $(A \otimes_K B)^\circ$ is solvable.

Proof. ad(i): This statement is straightforward to verify.

ad(ii): B° is Abelian, and thus by using part (i) for all $x, y \in A$ and $c, d \in B$ the statement $(*)$ $(x \otimes c) \circ (y \otimes d) = (x \circ y) \otimes (cd)$ is valid.
Let $T := A \otimes_K B$. Based on statement $(*)$ and a straightforward induction argument (see exercises) for all $m \in \mathbb{N}$ the condition $(T^\circ)^{(m)} \subseteq (A^\circ)^{(m)} \otimes_K B$ is true. A° is solvable, and thus part (ii) is now a consequence of part (i) of theorem 20.◇

In the next remark we analyze the Lie algebras $(K^{n \times n})^\circ$ (K a field, $n \in \mathbb{N}$) for being solvable. If the field possesses characteristic zero, then it is well-known that $(K^{n \times n})^\circ$ contains a simple K-subalgebra and is not solvable. We analyze the solvability of $(K^{n \times n})^\circ$ in arbitrary characteristic. For all $i, j \in \underline{n}$ let $e_{i,j}$ the matrix such that only the (i, j)-component is non-zero and, in addition, is exactly 1. The set of these matrices is the standard basis for $K^{n \times n}$.

Remark 16 (i) Let K be a field. $A := K^{2 \times 2}$ is a simple associative K-algebra. We prove that A° is solvable if and only if $char(K) \neq 2$ is valid. Let $B := \{e_{11}, e_{12}, e_{21}, e_{22}\}$ be the standard basis of A. We calculate

$$\begin{aligned}
e_{22} \circ e_{21} &= e_{21}, \\
e_{22} \circ e_{12} &= -e_{12}, \\
e_{22} \circ e_{11} &= 0_A, \\
e_{21} \circ e_{12} &= e_{22} - e_{11}, \\
e_{21} \circ e_{11} &= e_{21} \text{ and} \\
e_{12} \circ e_{11} &= -e_{12}.
\end{aligned}$$

Hence, we derive $(A^\circ)^{(1)} = \langle e_{21}, e_{12}, e_{22} - e_{11} \rangle_K$. In addition,

$$\begin{aligned}
e_{21} \circ e_{12} &= e_{22} - e_{11}, \\
e_{21} \circ (e_{22} - e_{11}) &= -2_K e_{21} \text{ and} \\
e_{12} \circ (e_{22} - e_{11}) &= -2_K e_{12}
\end{aligned}$$

are valid. If $char(K) = 2$ is true, then $(A^\circ)^{(2)} = \langle e_{22} - e_{11} \rangle_K$ and $(A^\circ)^{(3)} = \{0_A\}$ are valid. In the other case $(A^\circ)^{(1)} = (A^\circ)^{(2)}$ is true.

(ii) Let K be a field and $n \in \mathbb{N}_{\geq 3}$. $(K^{n \times n})^\circ$ is not solvable which will be proven now. For this, let $\{e_{ij} \mid 1 \leq i, j \leq n\}$ be the standard basis of $K^{n \times n}$. We have only to prove that $(K^{3 \times 3})^\circ$ is not solvable because $(K^{n \times n})^\circ$ contains a K-subalgebra isomorphic to $(K^{3 \times 3})^\circ$ (e.g. the set of those matrices $(a_{n,m})$ over K which are only non-zero for the components $a_{1,1}, a_{1,2}, a_{1,3}, a_{2,1}, a_{2,2}, a_{2,3}, a_{3,1}, a_{2,3}$ and $a_{3,3}$ are non-zero). We calculate

$$
\begin{aligned}
e_{11} \circ e_{12} &= e_{12}, \\
e_{11} \circ e_{13} &= e_{13}, \\
e_{11} \circ e_{31} &= -e_{31}, \\
e_{21} \circ e_{22} &= e_{21}, \\
e_{23} \circ e_{33} &= e_{23}, \\
e_{32} \circ e_{33} &= -e_{32}, \\
e_{12} \circ e_{21} &= e_{11} - e_{22} \text{ and} \\
e_{13} \circ e_{31} &= e_{11} - e_{33}.
\end{aligned}
$$

In addition,

$$
\begin{aligned}
e_{12} \circ e_{21} &= e_{11} - e_{22}, \\
e_{13} \circ e_{31} &= e_{11} - e_{33}, \\
e_{12} \circ e_{23} &= e_{13}, \\
e_{13} \circ e_{32} &= e_{12}, \\
e_{32} \circ e_{21} &= e_{31}, \\
e_{21} \circ e_{13} &= e_{23}, \\
e_{23} \circ e_{31} &= e_{21} \text{ and} \\
e_{31} \circ e_{12} &= e_{32}
\end{aligned}
$$

are valid. Thus, $\langle e_{12}, e_{13}, e_{21}, e_{23}, e_{31}, e_{32}, e_{11}-e_{22}, e_{11}-e_{33}\rangle_K \subseteq ((K^{3\times 3})^\circ)^{(n)}$ is proven for all $n \in \mathbb{N}$. We conclude that $(K^{3\times 3})^\circ$ is not solvable.

(iii) Let K be a field, $char(K) = 2$ and A a 4-dimensional central-simple associative unitary K-algebra. By using theorems in [25] a K-basis $B := \{1_A, i, j, k\}$ of A and elements $a \in K$, $b \in K \setminus \{0_K\}$ exist such that $i^2 + i = a1_A$, $j^2 = b1_A$ and $ij = k = j(i + 1_A)$ are valid. These K-algebras are quaternion algebras in characteristic 2. We prove that A° is solvable:

$$
\begin{aligned}
i \circ j &= ij + ji = j, \\
i \circ k &= ik + ki = i(ij) + (ji + j)i = (i + a1_A)j + j(i + a1_A) + ji = k \text{ and} \\
j \circ k &= jk + kj = j(ji + j) + ij^2 = j^2i + j^2 + ij^2 = bi + j^2 + bi = b1_A
\end{aligned}
$$

are valid. Thus, $(A^\circ)^{(1)} = \langle j, k, b1_A\rangle_K$, $(A^\circ)^{(2)} = \langle b1_A\rangle_K$ and $(A^\circ)^{(3)} = \{0_A\}$ are true.◇

Now we start executing the method mentioned before. For this, we prove the following lemma which is within this method the first step of an induction argument.

Lemma 6 Let K be a field and D a central-simple finite-dimensional associative K-division algebra. If $char(K) = 2$ is valid, then $dim_K(D) \neq 4$ is true. The following statements are equivalent:

(i) D is solvable.

(ii) $D = K1_A$

(iii) D° is solvable.

Proof. Because of $rad(D) = \{0\}$ and the central-simplicity of D the statements (i) and (ii) are equivalent. Proposition 4 lets us deduce that we have only to prove the implication from (iii) to (ii). Therefor, let part (iii) be valid and L be a maximal subfield of D. By using theorem 4.5.1 in [8] the isomorphism $D \otimes_K L \cong_{\mathcal{A}_1} L^{n \times n}$ and the statement $n = dim_K(L) = dim_L(D)$ are true. Because of remark 15 we deduce that $(L^{n \times n})^\circ$ is solvable as K-algebra. If $n = 1$ is true, then part (ii) is straightforward to prove. We assume $n \in \mathbb{N}_{\geq 2}$. If $n \neq 2$ would be valid, then the K-algebra $L^{n \times n}$ would contain a K-subalgebra T which is $\mathcal{A}_1$-isomorphic to $K^{3 \times 3}$. Hence, T° would be solvable as K-subalgebra of $(L^{n \times n})^\circ$. But this is a contradiction to the statements within remark 16. Let us assume $n = 2$. Only the case $char(K) \neq 2$ is to be analyzed. The K-algebra $L^{n \times n}$ contains a K-subalgebra T which is $\mathcal{A}_1$-isomorphic to $K^{2 \times 2}$. T° is solvable as K-subalgebra of $(L^{n \times n})^\circ$. Again, this a contradiction to the statements within remark 16. ⋄

Theorem 21 (Lie characterization of solvable associative algebras) Let K be a field, $char(K) \neq 2$ and A a finite-dimensional associative K-algebra. The following statements are equivalent:

(i) A is solvable.

(ii) A° is solvable.

Proof. Because of proposition 4 we only have to prove the implication from (ii) to (i). The corresponding proof is divided into the following four steps.

Step 1: Within this step we assume that A is a K-division algebra. We prove the statement by an induction argument based on the K-dimension $dim_K(A)$ of A. If A is one-dimensional, then the statement is valid. If $Z(A) = K1_A$ is true, then the proof is finished by using lemma 6. Thus we can assume that $K1_A \neq Z(A)$ is valid. A is a $Z(A)$-division algebra and $dim_{Z(A)}(A) \leq dim_K(A) - 1$ is true. A° is solvable as K-algebra, and thus it is solvable as $Z(A)$-algebra, too. By induction A is solvable, and hence A is a field.

Step 2: Within this step let A be simple. Thus, an element $n \in \mathbb{N}$ and a finite-dimensional associative K-division algebra D exist such that $A \cong_{\mathcal{A}_1} D^{n \times n}$ is valid. It is straightforward to prove that a K-subalgebra T of $D^{n \times n}$ exist which is $\mathcal{A}_1$-isomorphic to D (e.g. the set of diagonal matrices with entries over D which are identical in each component). T° is a K-subalgebra of $(D^{n \times n})^\circ$, and thus by applying step 1 the set D is a field. If $n \neq 1$ would be true, then $D^{n \times n}$ would possess a K-subalgebra $\mathcal{A}_1$-isomorphic to $K^{2 \times 2}$ (e.g. the set of those matrices $(a_{n,m})$ over K which are only non-zero for the components $a_{1,1}, a_{1,2}, a_{2,1}$ and $a_{2,2}$). But this algebra is not solvable based on remark 16. Hence, $n = 1$ is true and A is a field.

Step 3: Within this step let A be semisimple. Thus, a finite direct decomposition of A in simple K-ideals of A exists. Every K-ideal of this decomposition is a K-ideal of the solvable algebra A° and therefor Lie solvable, too. By using step 2 every K-ideal of this decomposition is commutative. We deduce that A is commutative.

Step 4: Now we prove the general case. Let A° be solvable. $A/rad(A)$ is a finite-dimensional semisimple associative K-algebra. In addition, $rad(A)^\circ$ is a K-ideal of A° and $A^\circ/rad(A)^\circ = (A/rad(A))^\circ$ is solvable. Hence, the general case is deductable by applying step 3.⋄

We will apply this characterization within this section to the tensor product of associative algebras. Therefor, we need the next proposition.

Proposition 5 *Let K be a field and A, B finite-dimensional associative K-algebras. If A and B are solvable, then $A \otimes_K B$ is solvable. If A and B are unitary, then the solvability of $A \otimes_K B$ implies the solvability of A and B.*

Proof. If A and B are unitary, then $A \otimes_K B$ contains subalgebras which are $\mathcal{A}_1$-isomorphic to A and B. In this case the solvability of $A \otimes_K B$ leads – based on part (iii) of corollary 3 – to the solvability of A and B.
Now let A and B be solvable and $T := A \otimes_K B$. Based on part (i) of remark 15 we deduce that $T \circ T \subseteq (A \circ A) \otimes_K B + A \otimes_K (B \circ B)$ is valid. A and B are solvable, and thus $A \circ A \subseteq rad(A)$ and $B \circ B \subseteq rad(B)$ are true. This statement implies $T \circ T \subseteq rad(A) \otimes_K B + A \otimes_K rad(B)$. The latter term is contained in $rad(T)$ because it is a nilpotent ideal of $A \otimes B$.⋄

This proposition is used to generalize part (ii) of remark 15.

Theorem 22 *(Solvability of the Lie tensor product) Let K be a field, $char(K) \neq 2$ and A, B associative finite-dimensional K-algebras. If A° and B° are solvable, then $(A \otimes_K B)^\circ$ is solvable.*

Solvable algebras

Proof. Let A° and B° be solvable. Theorem 21 lets us deduce that A and B are solvable. Based on proposition 5 we prove the solvability of $A \otimes_K B$. Again, by using theorem 21 the theorem is proven.◇

Counterexample 3 Let K be a field and $char(K) = 2$. Based on remark 16 we know that $K^{2\times 2}$ is solvable as Lie algebra. $K^{2\times 2} \otimes_K K^{2\times 2} \cong_{A_1} K^{4\times 4}$ is valid. The latter algebra is not Lie solvable by using remark 16.◇

3.2.2 A symmetric bilinear form

The second application of theorem 21 is done within this section. For this, we need some preliminary remarks and results. The aim is to characterize associative solvable algebras by using a special bilinear form derived from Cartan's criterion for the solvability of Lie algebras based on the Killing form.

Definition 7 Let A be an associative K-algebra. By $Nil(A)$ we denote the set of all nilpotent elements of A. If L is a Lie algebra, then for every $l \in L$ the multiplication with l is defined by $ad(l)$ - the so-called adjoint representation of l.◇

Definition and remark 8 Let K be a field and A a finite-dimensional associative K-algebra. We define

$$<,>_{\lambda,\rho}\colon A \times A \longrightarrow K, (a;b) \longmapsto tr(a\rho\, b\lambda + a\lambda\, b\rho).$$

$<,>_{\lambda,\rho}$ is a symmetric bilinear form on A, and the condition $rad(A) \subseteq Nil(A) \subseteq rad(<,>_{\lambda,\rho})$ is valid. (The radical $rad(f)$ of a symmetric bilinear form is the set of elements which are orthogonal to all other elements with respect to the bilinear form f.)

Proof. It is straightforward to verify that $<,>_{\lambda,\rho}$ is a bilinear form. $\rho\lambda = \lambda\rho$ is valid because of the associativity of A, and thus $<,>_{\lambda,\rho}$ is symmetric. Let $a \in Nil(A)$ and $b \in A$. $a\rho$ and $a\lambda$ are nilpotent. Again by using $\rho\lambda = \lambda\rho$ we conclude that $a\rho\, b\lambda$ and $a\lambda\, b\rho$ are nilpotent. Hence, $a \in rad(<,>_{\lambda,\rho})$ is valid (because the trace of a nilpotent endomorphism is zero).◇

Our aim is to characterize solvable associative algebras by using the bilinear form $<,>_{\lambda,\rho}$. In addition, we link the standard bilinear forms

$$<a,b>_\lambda := tr(a\lambda b\lambda) \text{ and}$$
$$<a,b>_\rho := tr(a\rho b\rho)$$

to the solvability of associative algebras. For this, we need the next definition: Let V be a K-space, B a K-basis of V and $\alpha \in End_K(V)$. By $M_B(\alpha)$

we denote the representative matrix of α with respect to B. The following theorem is due to Dickson (see [7]).[1]

---

[1] Leonard Eugene Dickson (January 22, 1874 to January 17, 1954) was an American mathematician. He was one of the first American researchers in abstract algebra, in particular the theory of finite fields and classical groups, and is also remembered for a three-volume history of number theory, History of the Theory of Numbers. Dickson considered himself a Texan by virtue of having grown up in Cleburne, where his father was a banker, merchant, and real estate investor. He attended the University of Texas at Austin, where George Bruce Halsted encouraged his study of mathematics. Dickson earned a B.S. in 1893 and an M.S. in 1894, under Halsted's supervision. Dickson first specialised in Halsted's own specialty, geometry. Both the University of Chicago and Harvard University welcomed Dickson as a Ph.D. student, and Dickson initially accepted Harvard's offer, but chose to attend Chicago instead. In 1896, when he was only 22 years of age, he was awarded Chicago's first doctorate in mathematics, for a dissertation titled The Analytic Representation of Substitutions on a Power of a Prime Number of Letters with a Discussion of the Linear Group, supervised by E. H. Moore. Dickson then went to Leipzig and Paris to study under Sophus Lie and Camille Jordan, respectively. On returning to the USA, he became an instructor at the University of California. In 1899 and at the extraordinarily young age of 25, Dickson was appointed associate professor at the University of Texas. Chicago countered by offering him a position in 1900, and he spent the balance of his career there. At Chicago, he supervised 53 Ph.D. theses; his most accomplished student was probably A. A. Albert. He was a visiting professor at the University of California in 1914, 1918, and 1922. In 1939, he returned to Texas to retire. Dickson married Susan McLeod Davis in 1902; they had two children, Campbell and Eleanor. Dickson was elected to the National Academy of Sciences in 1913, and was also a member of the American Philosophical Society, the American Academy of Arts and Sciences, the London Mathematical Society, the French Academy of Sciences and the Union of Czech Mathematicians and Physicists. Dickson was the first recipient of a prize created in 1924 by The American Association for the Advancement of Science, for his work on the arithmetics of algebras. Harvard (1936) and Princeton (1941) awarded him honorary doctorates. Dickson presided over the American Mathematical Society in 1917 to 1918. His December 1918 presidential address, titled 'Mathematics in War Perspective,' criticized American mathematics for falling short of those of Britain, France, and Germany: 'Let it not again become possible that thousands of young men shall be so seriously handicapped in their Army and Navy work by lack of adequate preparation in mathematics.' In 1928, he was also the first recipient of the Cole Prize for algebra, awarded annually by the AMS, for his book 'Algebren und ihre Zahlentheorie'. It appears that Dickson was a hard man: 'A hard-bitten character, Dickson tended to speak his mind bluntly; he was always sparing in his praise for the work of others. ... he indulged his serious passions for bridge and billiards and reportedly did not like to lose at either game.' 'He delivered terse and unpolished lectures and spoke sternly to his students. ... Given Dickson's intolerance for student weaknesses in mathematics, however, his comments could be harsh, even though not intended to be personal. He did not aim to make students feel good about themselves.' 'Dickson had a sudden death trial for his prospective doctoral students: he assigned a preliminary problem which was shorter than a dissertation problem, and if the student could solve it in three months, Dickson would agree to oversee the graduate student's work. If not the student had to look elsewhere for an advisor.' Dickson had a major impact on American mathematics, especially abstract algebra. His mathematical output consists of 18 books and more than 250 papers. The Collected Mathematical Papers of Leonard Eugene Dickson fill six large volumes. In 1901, Dickson published his first book 'Linear groups' with an exposition of the Galois field theory, a revision and expansion of his Ph.D. thesis. Teubner in Leipzig published the book, as there was no well-established American scientific publisher at the time. Dickson had already published

Lemma 2 *(radical of the standard trace form, Dickson 1923)* Let K be a field, $char(K) = 0$ and A a finite-dimensional associative unitary K-algebra. The statement

$$rad(<,>_\lambda) = rad(<,>_\rho) = rad(A)$$

is valid.

<u>Proof</u>. Let $a \in rad(A)$ and $b \in A$. $rad(A)$ is a K-ideal of A, and thus ab and ba are nilpotent. Hence, $(ab)\rho$ and $(ba)\lambda$ are nilpotent, and we conclude – because traces of nilpotent endomorphism are zero – $a \in rad(<,>_\lambda) \cap rad(<,>_\rho)$.
$<,>_\lambda$ and $<,>_\rho$ are associative bilinear forms of A. In particular, $rad(<$

---

43 research papers in the preceding five years; all but seven on finite linear groups. Parshall (1991) described the book as follows: 'Dickson presented a unified, complete, and general theory of the classical linear groups-not merely over the prime field $GF(p)$ as Jordan had done-but over the general finite field $GF(p^n)$, and he did this against the backdrop of a well-developed theory of these underlying fields. ... his book represented the first systematic treatment of finite fields in the mathematical literature.' An appendix in this book lists the non-Abelian simple groups then known having order less than 1 billion. He listed 53 of the 56 having order less than 1 million. The remaining three were found in 1960, 1965, and 1967. Dickson worked on finite fields and extended the theory of linear associative algebras initiated by Joseph Wedderburn and Cartan. He started the study of modular invariants of a group. In 1905, Wedderburn, then at Chicago on a Carnegie Fellowship, published a paper that included three claimed proofs of a theorem stating that all finite division algebras were commutative, now known as Wedderburn's theorem. The proofs all made clever use of the interplay between the additive group of a finite division algebra and its multiplicative group. Karen Parshall noted that the first of these three proofs had a gap not noticed at the time. Dickson also found a proof of this result but, believing Wedderburn's first proof to be correct, Dickson acknowledged Wedderburn's priority. But Dickson also noted that Wedderburn constructed his second and third proofs only after having seen Dickson's proof. She concluded that Dickson should be credited with the first correct proof. Dickson's search for a counterexample to Wedderburn's theorem led him to investigate non-associative algebras, and in a series of papers he found all possible three and four-dimensional (non-associative) division algebras over a field. In 1919 Dickson constructed Cayley numbers by a doubling process starting with real quaternions. His method was extended to a doubling of the real numbers to produce the complex numbers, and of the complex numbers to produce the real quaternions by A. A. Albert in 1922, and the procedure is known now as the Cayley-Dickson construction of composition algebras. Dickson proved many interesting results in number theory, using results of Vinogradov to deduce the ideal Waring theorem in his investigations of additive number theory. He proved the Waring's problem for $k = 7k \geq 7k \geq 7$ under the further condition of $(3k+1)/(2k-1) = [1.5k]+1(3^k+1)/(2^k-1) \leq [1.5^k] + 1(3^k+1)/(2^k-1) \leq [1.5^k]+1$ independently of Subbayya Sivasankaranarayana Pillai who proved it for $k = 6k \geq 6k \geq 6$ ahead of him. The three-volume 'History of the Theory of Numbers' (1919 to 1923) is still much consulted today, covering divisibility and primality, Diophantine analysis, and quadratic and higher forms. The work contains little interpretation and makes no attempt to contextualize the results being described, yet it contains essentially every significant number theoretic idea from the dawn of mathematics up to the 1920s except for quadratic reciprocity and higher reciprocity laws. A planned fourth volume on these topics was never written. A. A. Albert remarked that this three volume work 'would be a life's work by itself for a more ordinary man.'

$,>_\lambda)$ and $rad(<,>_\rho)$ are K-ideals of A. By using a theorem of Wedderburn we only need to prove that these ideals are nil. If $a \in rad(<,>_\rho)$, then $a^n \in rad(<,>_\rho)$ for all $n \in \mathbb{N}$ is valid. We deduce

$$tr((a^n)\rho 1\rho) = tr((a\rho)^n) = 0_K$$

for all $n \in \mathbb{N}$. We use exercise 42 on page 277 in [37] – and here $char(K) = 0$ is used – to conclude the nilpotency of $a\rho$. (Details for this exercise are mentioned within the exercises of this chapter.) Therefor, a is nilpotent, too. By using a similar argumentation we prove that $rad(<,>_\lambda)$ is nil.◇

Example 6 Let $A := \langle e, r \rangle_K$ be the algebra within example 5. $rad(A) = \langle r \rangle_K$ is valid, and a straightforward calculation leads to $rad(<,>_\rho) = rad(A)$. If $char(K) \neq 2$ is valid, then $rad(<,>_\lambda) = rad(A)$ is true. But within the case $char(K) = 2$ we calculate $A = rad(<,>_\lambda)$.◇

Theorem 23 *(characterization of solvable algebras by bilinear forms) Let K be a field, $char(K) = 0$ and A a finite-dimensional associative unitary K-algebra. The following statements are equivalent:*

(i) *A is solvable: $A \circ A \leq rad(A)$.*

(ii) *For all $a, b \in A$ the condition $a \circ b \in rad(<,>_\rho)$ is valid.*

(iii) *For all $a, b \in A$ the statement $a \circ b \in rad(<,>_\lambda)$ is valid.*

(iv) *For all $a \in A \circ A$ the condition $<a, a>_{\lambda,\rho} = <a^2, 1_A>_{\lambda,\rho}$ is true.*

(v) *For all $a, b, c \in A$ the statement $<a \circ b, c>_{\lambda,\rho} = <(a \circ b)c, 1_A>_{\lambda,\rho}$ is valid.*

Proof. Because of theorem 2 the parts (i), (ii) and (iii) are equivalent.

Now we prove that the statements (i) and (iv) are equivalent. By using theorem 21 part (i) is equivalent to the solvability of A°. Cartan's criteria for solvable Lie algebras is equivalent to the statement that for all $a \in A \circ A$ the condition $tr(ad(a)^2) = 0_K$ is valid. If $a \in A$ is true, then

$$\begin{aligned} ad(a) &= a\rho - a\lambda \text{ and} \\ ad(a)^2 &= a^2\rho + a^2\lambda - 2a\rho a\lambda \end{aligned}$$

are valid. Thus, the parts (i) and (iv) are equivalent.

Now we focus on the implication from (i) to (v). For this, let A be solvable. $A \circ A \subseteq rad(A)$ is valid by definition, and thus $a \circ b, (a \circ b)c \in rad(A)$ is true for all $a, b, c \in A$. By using theorem 2 part (v) is proven.

Solvable algebras 107

Let part (v) be valid. By using the criteria of Cartan for solvable Lie algebras and theorem 21 we only have to prove that $A \circ A$ is contained in the radical of the Killing form.[2] Let $a, b, c \in A$. We have to prove $tr(ad(a \circ b)ad(c)) = 0_K$. $ad(a \circ b) = ad(a) \circ ad(b)$ is valid. In addition,

$$ad(a)ad(b)ad(c) =$$
$$(a\rho - a\lambda)(b\rho - b\lambda)(c\rho - c\lambda) =$$
$$((ab)\rho - a\rho b\lambda - a\lambda b\rho + (ba)\lambda)(c\rho - c\lambda) =$$
$$(abc)\rho - (ab)\rho c\lambda - a\rho b\lambda c\rho + a\rho b\lambda c\lambda - a\rho b\lambda c\lambda + a\lambda b\rho c\lambda + (ba)\lambda c\rho - (cba)\lambda$$

is true. By a similar calculation

$$ad(b)ad(a)ad(c) =$$
$$(b\rho - b\lambda)(a\rho - a\lambda)(c\rho - c\lambda) =$$
$$((ba)\rho - b\rho a\lambda - b\lambda a\rho + (ab)\lambda)(c\rho - c\lambda) =$$
$$(bac)\rho - (ba)\rho c\lambda - b\rho a\lambda c\rho + b\rho a\lambda c\lambda - b\rho a\lambda c\lambda + b\lambda a\rho c\lambda + (ab)\lambda c\rho - (cab)\lambda$$

---

[2]Wilhelm Karl Joseph Killing (10 May 1847 to 11 February 1923) was a German mathematician who made important contributions to the theories of Lie algebras, Lie groups, and non-Euclidean geometry. Killing studied at the University of Münster and later wrote his dissertation under Karl Weierstrass and Ernst Kummer at Berlin in 1872. He taught in gymnasia (secondary schools) from 1868 to 1872. He became a professor at the seminary college Collegium Hosianum in Braunsberg (now Braniewo). He took holy orders in order to take his teaching position. He became rector of the college and chair of the town council. As a professor and administrator Killing was widely liked and respected. Finally, in 1892 he became professor at the University of Münster. Killing and his spouse had entered the Third Order of Franciscans in 1886. In 1878 Killing wrote on space forms in terms of non-Euclidean geometry in Crelles Journal, which he further developed in 1880 as well as in 1885. Recounting lectures of Weierstrass, he there introduced the hyperboloid model of hyperbolic geometry described by Weierstrass coordinates. He is also credited with formulating transformations mathematically equivalent to Lorentz transformations in n dimensions in 1885. Killing invented Lie algebras independently of Sophus Lie around 1880. Killings university library did not contain the Scandinavian journal in which Lies article appeared. (Lie later was scornful of Killing, perhaps out of competitive spirit and claimed that all that was valid had already been proven by Lie and all that was invalid was added by Killing.) In fact Killings work was less rigorous logically than Lies, but Killing had much grander goals in terms of classification of groups, and made a number of unproven conjectures that turned out to be true. Because Killings goals were so high, he was excessively modest about his own achievement. From 1888 to 1890, Killing essentially classified the complex finite-dimensional simple Lie algebras, as a requisite step of classifying Lie groups, inventing the notions of a Cartan subalgebra and the Cartan matrix. He thus arrived at the conclusion that, basically, the only simple Lie algebras were those associated to the linear, orthogonal, and symplectic groups, apart from a small number of isolated exceptions. Elie Cartans 1894 dissertation was essentially a rigorous rewriting of Killings paper. Killing also introduced the notion of a root system. He discovered the exceptional Lie algebra g2 in 1887; his root system classification showed up all the exceptional cases, but concrete constructions came later. As A. J. Coleman says 'He exhibited the characteristic equation of the Weyl group when Weyl was 3 years old and listed the orders of the Coxeter transformation 19 years before Coxeter was born.'

is valid. We deduce

$$tr((ad(a) \circ ad(b))(ad(c))) =$$
$$< b \circ a, c >_{\lambda,\rho} + tr((abc)\rho - (bac)\rho) + tr((cab)\lambda - (cba)\lambda) =$$
$$< b \circ a, c >_{\lambda,\rho} + < (a \circ b)c, 1_A >_{\lambda,\rho} .$$

By using our assumptions the last term within this calculation is exactly 0_K. ⋄

Corollary 5 *Let K be a field, $char(K) = 0$ and A a finite-dimensional associative unitary K-algebra. If $<,>_{\lambda,\rho}$ is associative, then A is solvable.*

Proof. The proof is a direct consequence of part (iv) of theorem 23. ⋄

Counterexample 4 Let A be the algebra of example 1. We calculate $< e, e >_{\lambda,\rho} = 2_K$ and $< e^2, 1_A >_{\lambda,\rho} = 3_K$. Hence, A is solvable but the bilinear form $<,>_{\lambda,\rho}$ is not associative. The problem of determining those associative algebras for which $<,>_{\lambda,\rho}$ is associative remains open. ⋄

Before applying theorem 21 to certain examples of algebras we transfer theorem 23 to non-unitary algebras. For this transfer we need the following remark about the adjunction of an unit.

Remark 17 *Let K be a field and A a finite-dimensional associative K-algebra. The following statements are valid:*

(i) A is solvable if and only of (K, A) is solvable.

(ii) For all $a, b \in A, k, l \in K$ the condition $(k; a) \circ (l; b) = (0_K; a \circ b)$ is true. In particular, $(K, A) \circ (K, A) = \{0_K\} \times (A \circ A)$ is valid.

(iii) If $a, b \in A$, then $tr(a\rho) = tr((0_K; a)\rho)$, $tr(a\lambda) = tr((0_K; a)\lambda)$ and $tr(a\rho b\lambda) = tr((0_K; a)\rho (0_K; b)\lambda)$ are valid.

Proof. ad(i): This part is a direct consequence of remark 7, theorem 19 and corollary 3.

ad(ii): This part can be verified by a basic calculation.

ad(iii): If B is a K-basis of A, then $(\{0_K\} \times B) \cup \{(1_K; 0_A)\}$ is a K-basis of (K, A). By using this basis we can prove part (iii) by a straightforward calculation. ⋄

Theorem 24 *Let K be a field, $char(K) = 0$ and A a associative finite-dimensional K-algebra. The following statements are equivalent:*

Solvable algebras   109

(i) A is solvable.

(ii) For all $x \in A \circ A$ the conditions $tr(x\rho) = 0_K$ and $x \in rad(<,>_\rho)$ are valid.

(iii) For all $x \in A \circ A$ the statements $tr(x\lambda) = 0_K$ and $x \in rad(<,>_\lambda)$ are true.

(iv) For all $a \in A \circ A$ the condition $2 \cdot tr(a\lambda a\rho) = tr(a^2 \lambda a^2 \rho)$ is true.

(v) For all $a, b, c \in A$ the statement $tr(((a \circ b)c)\lambda + ((a \circ b)c)\rho) = <a \circ b, c>_{\lambda,\rho}$ is valid.

Proof. We prove that the statements are equivalent to the corresponding ones within theorem 23.

(i) $\iff$ (ii): The following argumentation is valid:

A solvable $\iff_{remark\,17}$
(K, A) solvable $\iff_{theorem\,23}$
$(K, A) \circ (K, A) \subseteq rad(<,>_\rho) \iff_{remark\,17}$
$\forall a, b, c \in A, k \in K : tr((0_K; a \circ b)\rho(k; c)\rho) = 0_K \iff_{remark\,17}$
$\forall a, b, c \in A, k \in K : tr(((a \circ b))c\rho) + k \cdot tr((a \circ b)\rho) = 0_K$.

By differentiating between the cases $k = 0_K$ and $c = 0_A, k = 1_K$ we deduce the equivalence of the parts (i) and (ii). A similar argumentation lets us deduce the equivalence of the statements (i) and (iii).

(i) $\iff$ (iv): The following argumentation is valid:

A solvable $\iff_{remark\,17}$
(K, A) solvable $\iff_{theorem\,23}$
$\forall a \in (K, A) \circ (K, A) :<a, a>_{\lambda,\rho} = <a^2, 1_{(K,A)}>_{\lambda,\rho} \iff_{remark\,17}$
$\forall a \in (\{0_K\} \times A) \circ (\{0_K\} \times A) :<a, a>_{\lambda,\rho} = <a^2, 1_{(K,A)}>_{\lambda,\rho} \iff_{remark\,17}$
$\forall a \in A \circ A : 2 \cdot tr(a\lambda a\rho) = tr(a^2 \lambda a^2 \rho)$.

(i) $\iff$ (v): The following argumentation is valid:

A solvable $\iff_{remark\,17}$
(K, A) solvable $\iff_{theroem\,23}$
$\forall a, b, c \in A, k \in K :<(0_K; a \circ b), (k; c)>_{\lambda,\rho} =$
$= <(0_K; a \circ b)(k; c), (1_K; 0_K)>_{\lambda,\rho} \iff_{remark\,17}$
$\forall a, b, c \in A, k \in K : tr(((a \circ b)c + k(a \circ b))(\lambda + \rho)) =$
$= <a \circ b, c>_{\lambda,\rho} + tr(k(a \circ b)(\lambda + \rho)) \iff$
$\forall a, b, c \in A : tr((a \circ b)c(\lambda + \rho)) = <a \circ b, c>_{\lambda,\rho}.\diamond$

Example 7 Let $A = \langle e, r \rangle_K$ be the algebra within example 5. We calculate $A \circ A = \langle r \rangle_K$. Let $k \in K$. $(kr)^2 = 0_A$ and $tr(((kr)^2)\lambda((kr)^2)\rho) = 0_K$ are valid. $(kr)\rho(kr)\lambda$ is the zero-function on A. By using theorem 24 we conclude that in the case $char(K) = 0$ the algebra A is solvable.⋄

Now we will focus on examples for the Lie characterization within theorem 21. For this, we study group algebras and the algebra of lower triangular matrices over a field.

3.2.3 Group algebras

Within this subsection let K be a field and G a finite group. In [32] it was proven that $(KG)^\circ$ is solvable if and only if G is Abelian, the derived subgroup of G is a p-group (in the case $p := char(K) \notin \{0, 2\}$, G non-Abelian) or G possesses a subgroup of index 2 with a derived 2-subgroup (in the case $char(K) = 2$, G non-Abelian). Four years later it was proven in [33] on what terms $E(KG)$ is solvable: G is Abelian or $G/O_p(G)$ is Abelian (in the case $p := char(K) \neq 0$, G non-Abelian). $O_p(G)$ is the so-called p-core of G which is the greatest p-normal subgroup of G. By using Sylow's theorems G' is a p-group if and only if $G/O_p(G)$ is Abelian. Thus, in the case $char(K) \neq 2$ th Lie algebra $(KG)^\circ$ is solvable if and only if $E(KG)$ is solvable. This result is generalized by theorem 5.4 in [3] and theorem 21. If $char(K) = 2$ is valid, then the solvability of $E(KG)$ implies the solvability of $(KG)^\circ$ (see theorem 25 and proposition 4). But if we focus on the group S_3 within the case $char(K) = 2$, then $(KG)^\circ$ is and $E(KG)$ is not solvable. Now we analyze on what terms KG is solvable. By $Aug(KG)$ we denote the so-called augmentation ideal of KG which is the kernel of the linearization to KG of the function $G \longrightarrow K, g \mapsto 1_K$.

Theorem 25 *(solvability of group algebras) KG is solvable if and only if $E(KG)$ is solvable.*

Proof. In the case $char(K) = 0$ the statement is deductable from theorem 5.4 in [3]. Let $p := char(K) \neq 0$. If KG is solvable, then $E(KG)$ is solvable, too (see again theorem 5.4 in [3]). Let $E(KG)$ be solvable. If G is Abelian, then KG is commutative and therefor solvable, too. Let G be non-Abelian. As described within the introduction to section 3.2.3 the derived subgroup G' is a p-group. By using proposition 1.2 in [22] we deduce that $KG/(KG\,Aug(KG')) \cong_{A_1} K(G/G')$ is valid. Hence, $KG/(KG\,Aug(KG'))$ is commutative. We have only to prove that $KG\,Aug(KG')$ is nilpotent. Corollary 4.7 in [35] let us conclude that $Aug(KG')$ is nilpotent. $KG\,Aug(KG') = Aug(KG')\,KG$ is valid (see introduction before proposition 1.2 in [22]). We conclude that with $Aug(KG')$ also $KG\,Aug(KG')$ is nilpotent.⋄

Until today it is not known how to calculate the classes for solvability for the

Solvable algebras 111

three structures KG, $(KG)^\circ$ and $E(KG)$. Within the exercises the reader may bound the class of solvability of these structures for a group with cyclic derived subgroup. The next section is dedicated to the determination of these classes for the algebra of lower triangular matrices.

3.2.4 Triangular matrices

Proposition 6 *Let A be a finite-dimensional associative K-algebra. The following statements are valid:*

(i) *For all $n \in \mathbb{N}$ the condition $(A^\circ)^{(n)} \subseteq A^{(n)}$ is true. If A is solvable, then for all $n \in \mathbb{N}$ the statement $A^{(n)} \subseteq rad(A)^{<2^{n-1}>}$ is true. If A is solvable, then A° is solvable and*

$$st(A^\circ) \leq st(A) \leq min\{n \in \mathbb{N} \mid 2^{n-1} \geq cl(rad(A))\} = \lfloor log_2(cl(rad(A))) \rfloor$$

is true.

(ii) *If for all $n \in \mathbb{N}$ the set $(A^\circ)^{(n)}$ is an ideal of A, then for all $n \in \mathbb{N}$ the identities $(A^\circ)^{(n)} = A^{(n)}$ and $st(A^\circ) = st(A)$ are valid.*

(iii) *Let A be unitary and solvable. For all $n \in \mathbb{N}$ the statement $E(A)^{(n)} \subseteq 1_A + A^{(n)}$ is true. If $E(A)$ is solvable, then $st(E(A)) \leq st(A)$ is valid.*

Proof. ad (i): The first inclusion is true by definition of $A^{(n)}$. Let A be solvable. Thus, $A \circ A \subseteq rad(A)$ is valid. $rad(A)$ is a K-ideal of A, and thus we deduce $A^{(1)} \subseteq rad(A)$. For all $n \in \mathbb{N}$ the factor algebra $rad(A)^{<2^{n-1}>}/rad(A)^{<2^n>}$ is commutative and $rad(A)^{<2^n>}$ is a K-ideal of A. We conclude that part (i) is true.

ad(ii): This statement is straightforward to prove by definition of $(A^\circ)^{(n)}$ and $A^{(n)}$.

ad(iii): $A^{(1)}$ is a K-ideal of $rad(A)$, and thus $1_A + A^{(1)}$ is a normal subgroup of $1_A + rad(A)$. Let $x, y \in E(A)$. Because of $[x, y] = 1_A + x^{-1}y^{-1}(x \circ y)$ we deduce $E(A)^{(1)} \subseteq 1_A + A^{(1)}$. We use an induction argument to prove for all $n \in \mathbb{N}$ that $1_A + A^{(n)}$ is a subgroup of $1_A + rad(A)$ and $1_A + A^{(n+1)}$ is a normal subgroup possessing an Abelian factor group $(1_A + A^{(n)})/(1_A + A^{(n+1)})$. $A^{(n+1)}$ is a K-subalgebra of $A^{(n)}$, and thus $1_A \in 1_A + A^{(n+1)}$ and $(1_A + a)(1_A + b) \in 1_A + A^{(n+1)}$ are valid for all $a, b \in A^{(n+1)}$. If $a \in A^{(n+1)}$ is true, then $1_A + a$ possesses an inverse element in $1_A + A^{(n)}$. Hence, $a' \in A'$ exists such that $(1_A + a)^{-1} = 1_A + a'$ is true. Because of $1_A = (1_A + a')(1_A + a)$ we deduce $a' = -a - a'a$. $A^{(n+1)}$ is a K-ideal of $A^{(n)}$, and thus $a' \in A^{(n+1)}$ is true. For an element $b \in A^{(n)}$ we use theorem 3 to deduce

$$(1_A + a')(1_A + b)(1_A + a) = 1_A + a'b + b + ba + a'ba.$$

$A^{(n+1)}$ is a K-ideal of $A^{(n)}$. Therefor $1_A + A^{(n+1)}$ is a normal subgroup of $1_A + A^{(n)}$. Finally, for all $x, y \in A^{(n)}$ the commutator-identity

$$[1_A + x, 1_A + y] =$$
$$1_A + (1_A + x')(1_A + y')((1_A + x) \circ (1_A + y)) =$$
$$1_A + (1_A + x')(1_A + y')(x \circ y)$$

is true. By definition of $A^{(n+1)}$ we deduce that the relevant factor group is Abelian.◇

Calculation 1 *(classes of solvability for triangular matrices)* Let K be a field, $n \in \mathbb{N}$ and $A := \Delta_{u,n}$. A, A° and $E(A)$ are solvable (see section 1.3.2 and proposition 6). Within this section we calculate the classes of solvability for these three structures. For the associative algebra A and the Lie algebra A° the following result is valid:

For all $n \in \mathbb{N}$ the identity

$$(A^\circ)^{(n)} = A^{(n)} = rad(A)^{<2^{n-1}>}$$

is true. In particular, proposition 6 lets us deduce for the minimal m such that $2^{m-1} \geq cl(rad(A)) = n$ is valid the classes of solvability of the two structures is identical to

$$st(A) = st(A^\circ) = m = \lfloor log_2(cl(rad(A))) \rfloor + +1.$$

For the proof let $B := \{e_{ij} \mid 1 \leq j \leq i \leq n\}$ be the standard basis of A. We want to prove that for all $r \in \mathbb{N}$ the identities

$$rad(A)^{<r>} \circ rad(A)^{<r>} = rad(A)^{<2r>} \text{ and}$$
$$A \circ A = rad(A)$$

are valid. Let $r \in \mathbb{N}$. It is well-known that

$$rad(A)^{<r>} = \langle e_{ij} \mid r+1 \leq i \leq n, 1 \leq j \leq i-r \rangle_K$$

is true. For all $1 \leq i, j \leq n$ such that $i \neq j$ is valid the identity $e_{ij} = e_{i1} \circ e_{1j}$ is true. We deduce $A \circ A = rad(A)$. Straightforward to prove is the identity $rad(A)^{<r>} \circ rad(A)^{<r>} \subseteq rad(A)^{<2r>}$. Let $i, j \in \underline{n}$ such that $2r+1 \leq i \leq n$ and $1 \leq j \leq i - 2r$ are true. We calculate $e_{(r+j)j} \circ e_{i(r+j)} = e_{ij}$. We have only to prove that $e_{(r+j)j}$ and $e_{i(r+j)}$ are contained in $rad(A)^{<r>}$. Because of $1 \leq j \leq i - 2r$ we deduce $1 \leq j \leq n - r$ and $r + 1 \leq r + j \leq n$. $1 \leq j \leq (r+j) - j$ is valid. In addition, $r + 1 \leq 2r + 1 \leq i \leq n$ is true. Finally, $1 \leq j \leq i - 2r$ is valid, and therefor $1 \leq r + 1 \leq r + j \leq i - r$ is true.

Solvable algebras

Now we focus on the unit group of A. Theorem 1 of page 125 in [39] lets us deduce that for all $n \in \mathbb{N}_{\geq 2}$ the identity

$$st(1_A + rad(A)) = [log_2(n-1)] + 1$$

is valid. In the case $n = 1$ it is straightforward to prove $st(1_A + rad(A)) = 1$. First, we do a calculation to link the previous results:

($*$) $\forall n \in \mathbb{N}_{\geq 2} : min\{m \in \mathbb{N} \mid 2^{m-1} \geq n\} = [log_2(n-1)] + 2$.

Let $n \in \mathbb{N}_{\geq 2}$, $x := [log_2(n-1)]$ and $m := min\{m \in \mathbb{N} \mid 2^{m-1} \geq n\}$. x is the unique integer such that $x \leq log_2(n-1) < x+1$ is valid. Hence, $2^x \leq n-1 < 2^{x+1}$ is true, and thus $2^{x+2-1} = 2^{x+1} > n-1$ is valid. We deduce $2^{x+1} \geq n$ and $x+2 \geq m$. Let us assume $x+2 \geq m+1$. $x \geq m-1$ and $2^x \geq 2^{m-1} \geq n$ would be valid, and hence also $x \geq log_2(n) > log_2(n-1) \geq x$. This is a contradiction and ($*$) is valid.

For the derived subgroup of $E(A)$ the following result is valid:

($**$) If $\mid K \mid \neq 2$ is valid, then $E(A)' = 1_A + rad(A)$ is true.

Thus, in the case $\mid K \mid \neq 2$ the classes of solvability for $E(A)$, A and of $A°$ are identical. If $\mid K \mid = 2$ is valid, then it is straightforward to prove that $E(A) = 1_A + rad(A)$ is true. In this case for $n \neq 1$ resp. for $n = 1$ the identity $st(E(A)) = st(A) - 1$ resp. $st(A) = st(E(A)) = 1$ is valid.

Now we prove the statement ($**$). This will be done by a calculation divided into several steps.

(i) If for all $i \in \underline{n}$ the condition $a_{ii} \in K \setminus \{0_K\}$ is valid, then $(\sum_{i=1}^{n} a_{ii} e_{ii})^{-1} = \sum_{i=1}^{n} a_{ii}^{-1} e_{ii}$ is true.

(ii) Let $2 \leq i \leq n$ and $1 \leq j \leq i-1$. We calculate $(1_A + e_{ij})(1_A - e_{ij}) = 1_A$.

(iii) Let $2 \leq i \leq n$, $1 \leq j \leq i-1$ and $a_{kk} \in K \setminus \{0_K\}$.

Based on parts (i) and (ii) the following calculation is valid:

$$[1_A + e_{ij}, \sum_{k=1}^{n} a_{kk}e_{kk}] =$$

$$(1_A - e_{ij})(\sum_{k=1}^{n} a_{kk}^{-1}e_{kk})(1_A + e_{ij})(\sum_{k=1}^{n} a_{kk}e_{kk}) =$$

$$(\sum_{k=1}^{n} a_{kk}^{-1}e_{kk} - \sum_{k=1}^{n} a_{kk}^{-1}e_{ij}e_{kk}) \cdot (\sum_{k=1}^{n} a_{kk}e_{kk} + \sum_{k=1}^{n} a_{kk}e_{ij}e_{kk}) =$$

$$(\sum_{k=1}^{n} a_{kk}^{-1}e_{kk} - a_{jj}^{-1}e_{ij})(\sum_{k=1}^{n} a_{kk}e_{kk} + a_{jj}e_{ij}) =$$

$$1_A + \sum_{k=1}^{n} a_{jj}a_{kk}^{-1}e_{kk}e_{ij} - \sum_{k=1}^{n} a_{jj}^{-1}a_{kk}e_{ij}e_{kk} =$$

$$1_A + a_{ii}^{-1}a_{jj}e_{ij} - a_{jj}^{-1}a_{jj}e_{ij} =$$

$$1_A + (a_{ii}^{-1}a_{jj} - 1_K)e_{ij}.$$

(iv) Let $d \in K$, $2 \leq i \leq n$ and $1 \leq j \leq i-1$. If $d \neq -1_K$, $a_{jj} := d + 1_K$ and $a_{kk} := 1_K$ for all $k \in \underline{n}$ and $k \neq j$, then part (iii) lets us deduce that $1_A + d\,e_{ij} \in E(A)'$ is valid. Let $d = -1_K$. If $char(K) \neq 2$ is true, then we have already proven $1_A + e_{ij} \in E(A)'$. Part (ii) is used to verify $1_A - e_{ij} \in E(A)'$. Let $char(K) = 2$. We choose an element $a \in K$ such that $0_K \neq a \neq 1_K$ is valid. For this element $0_K \neq a + 1_K \neq 1_K$ is valid. It was already proven that $1_A + a\,e_{ij}$ and $1_A + (a + 1_K)\,e_{ij}$ are contained in $E(A)'$. Hence, $(1_A + a\,e_{ij})(1_A + (a+1_K)\,e_{ij}) = 1_A + e_{ij}$ is contained in $E(A)'$.

(v) Let $M \in rad(A)$. We calculate $1_A + M = 1_A + \sum_{i=2}^{n}\sum_{j=1}^{i-1} m_{ij}e_{ij}$. This term can be represented as a product (which may be proven by the reader as an exercise) $\prod_{i=2}^{n}\prod_{j=1}^{i-1}(1_A + m_{ij}e_{ij})$. This representation is the key idea for linking the additive and the multiplicative structure:

$$1_A + M = 1_A + \sum_{i=2}^{n}\sum_{j=1}^{i-1} m_{ij}e_{ij} = \prod_{i=2}^{n}\prod_{j=1}^{i-1}(1_A + m_{ij}e_{ij}).$$

Based on part (iv) we deduce $1_A + rad(A) \subseteq E(A)'$. The opposite inclusion is true by using parts (i) and (iii) of proposition 6.

The next table shows the classes for solvability of the three structures related to algebras of triangular matrices for $n \in \underline{17}$.

Solvable algebras

n	$st(A) = st(A^\circ)$	$st(E(A))$	$st(E(A))$ over GF(2)
1	1	1	1
2	2	2	1
3	3	3	2
4	3	3	2
5	4	4	3
.	.	.	.
8	4	4	3
9	5	5	4
.	.	.	.
16	5	5	4
17	6	6	5 ⋄

3.3 Compatibilities

Remark 18 *Let A be an associative commutative K-algebra. The following statements are valid:*

(i) If A is semisimple, then every right artian subalgebra is semisimple.

(ii) Let K be a field. If A is separable, then every subalgebra of A is separable.

Proof. ad(i): Let T be a right artian K-subalgebra of A. A is commutative and semisimple, and thus $rad(T) = rad(A) \cap T = \{0_A\}$ is valid. T is right artian with radical zero and therefor semisimple.

ad(ii): A is separable, and based on part (ii) of theorem 1 the tensor product $A \otimes_K A$ is semisimple and A is finite-dimensional. T is a K-subalgebra of A. In particular, T is finite-dimensional. $T \otimes_K T$ is a K-subalgebra of $A \otimes_K A$. Part (i) lets us deduce that $T \otimes_K T$ is semisimple. Based on part (ii) of theorem 1 we have only to prove that T is unitary. Part (i) of corollary 1 is used to prove that A is semisimple, and by using part (i) the subalgebra T is semisimple. The proof is finished by applying theorem 1. ⋄

The next two results demonstrate compatibility properties of solvable associative algebras. If A is an associative K-algebra and $a \in A$ algebraical over K, then we define $min_{a,K}$ and $char_{a,K}$ to be the minimal and characteristical polynomial of a over K.

Theorem 26 *(separability of subalgebras of solvable associative algebra) Let A be a finite-dimensional solvable associative K-algebra. The following statements are valid:*

(i) *If T is a K-subalgebra of A, then $rad(T) = rad(A) \cap T$ is valid and T is solvable. In particular, $rad(A)$ is exactly the set of nilpotent elements of A and the unique maximal nilpotent K-subalgebra of A.*

(ii) *Let $A/rad(A)$ separable. If T is a K-subalgebra of A, then $T/rad(T)$ is separable. In particular, every semisimple K-subalgebra of A is separable.*

(iii) *Let $n \in \mathbb{N}$. If K is a splitting field for A, then it is a splitting field for every K-subalgebra of A. In particular, every K-subalgebra T of K^n is $\mathcal{A}_1$-isomorphic to $K^{dim_K(T)}$.*

Proof. ad(i): $rad(A) \cap T$ is a nilpotent ideal of T. Based on well-known isomorphism theorems $T/(rad(A) \cap T)$ is isomorphic to a K-subalgebra of $A/rad(A)$. Part (i) of remark 18 is used to prove that $T/(rad(A) \cap T)$ is semisimple. In particular, $rad(T) = rad(A) \cap T$ is valid and T is solvable. Let r be a nilpotent element of A. $\langle r \rangle_A = \langle r, r^2, ..., r^{cl(r)-1} \rangle_K$ is valid. Within this finite-dimensional commutative K-subalgebra the element r is nilpotent. Based on the result proven so far we deduce $r \in rad(A)$.

ad(ii): As deduced in part (i) the factor algebra $T/(rad(A) \cap T)$ is isomorphic to a K-subalgebra of $A/rad(A)$. Based on part (ii) of remark 18 we conclude that $T/(rad(A) \cap T)$ is separable. Part (i) shows us $rad(T) = rad(A) \cap T$, and thus part (ii) is proven.

ad(iii): The proof is executed within four steps. Let $B := \{e_1, ..., e_n\}$ be the standard basis of K^n.

(1) Let T be a K-subalgebra of A. Based on part (i) the algebra $T/rad(T)$ is isomorphic to a subalgebra of $A/rad(A)$. Hence, we need only to show the add-on of part (iii).

(2) We deduce now that the add-on is to be proven only for those subalgebras containing 1_{K^n}. Let S be a K-subalgebra of K^n. Based on part (i) of remark 18 the K-subalgebra S is semisimple. Theorem 1 lets us deduce that S is unitary. 1_S is an idempotent element of K^n, and thus a subset X of $\underline{n}$ exists such that $1_S = \sum_{x \in X} e_x$ is valid. If $s \in S$, presented as $s = \sum_{i=1}^{n} k_i e_i$, then $1_S = s1_S = \sum_{x \in X} k_x e_x$ is true. Therefore, S is contained in $\langle e_x \mid x \in X \rangle_K$ as a K-subalgebra. Within this K-ideal of K^n the element 1_S is the unit element. In addition, this K-ideal is isomorphic to $K^{|X|}$. Hence, we can assume that S contains the unit element of K^n.

(3) Let S be a K-subalgebra of K^n which is a field. Based on part (2)

we can assume that S contains the unit element of K^n. Let $a \in A$. a can be presented as $a = \sum_{i=1}^{n} k_i e_i$. $char_{M_B(a\rho),K} = \prod_{i=1}^{n}(t - k_i)$ is valid. ρM_B is an algebra monomorphism of A into $K^{n \times n}$, and thus $min_{a,K} = min_{M_B(a\rho),K} \mid char_{M_B(a\rho),K}$ is true. Hence, $min_{a,K}$ is splitting over K. S is a field, and thus the minimal polynomial of every element of S is irreducible over K. We have proven that S coincide with $K \cdot 1_A$.

(4) Let S be a K-subalgebra of K^n. Based on part (i) of remark 18 the algebra S is semisimple. Thus, S possesses a direct decomposition into ideals of S which are fields. By applying part (3) we finish the proof.◇

Theorem 27 *Let K be a field and A a finite-dimensional associative solvable K-algebra possessing a separable factor algebra by its nilradical. If T is a K-subalgebra of A, then T possesses a radical complement contained in a radical complement of A. If T is commutative, then for every algebra complement D of $rad(A)$ in A and every algebra complement C of $rad(T)$ in T the condition $D \cap T \subseteq C$ is valid.*

Proof. Let T be a K-subalgebra of A. Based on parts (i) and (ii) of theorem 26 we deduce that $rad(T) = rad(A) \cap T$ is valid and that T is possessing a separable factor algebra by its nilradical. Theorem 13 implies that a radical complement of C of T exists. The next step is to deduce:

(∗) For every radical complement D of A an element $r \in rad(T)$ exists such that $(D \cap T)^{(r)} \subseteq C$ is valid.

Let D be a radical complement of A. $D \cap T$ is a K-subalgebra of the commutative separable algebra D. Based on part (ii) of remark 18 we deduce that $D \cap T$ is a separable K-subalgebra of T. Now we can end the proof for (∗) by using part (i) of corollary 4.
In addition, C is a separable K-subalgebra of A. Based on part (i) of corollary 4 an element $x \in rad(A)$ exists such that $C \subseteq D^{(x)} \cap T$ is true. Part (ii) of corollary 4 lets us deduce that $D^{(x)}$ is a radical complement of A. Now we use (∗) and part (ii) of corollary 4 to prove $C = D^{(x)} \cap T$. The second part of the theorem is also deductable from (∗).◇

Corollary 6 *(radical complements of sub- and factor algebras of commutative associative algebras) Let K be a field, A a finite-dimensional commutative K-algebra possessing a separable factor algebra by its nilradical, T a K-subalgebra and I an K-ideal of A. If D is the algebra complement of $rad(A)$ in A, then $T \cap D$ resp. $(D + I)/I$ is the algebra complement of $rad(T)$ resp. of $rad(A/I)$ in T resp. in A/I.*

Proof. This corollary is a consequence of part (ii) of theorem 18, theorem 27 and part (vi) of remark 16.◇

Now we focus on examples showing that the intersection of radical complements with subalgebras is not compatible in general: the resulting subalgebra may not be a radical complement of the subalgebra.

Example 8 We focus on the K-algebra $A := \Delta_{u,4}$ over an arbitrary field K (see section 1.3.2). The nilradical of A is exactly the K-subalgebra of strict lower triangular matrices over K, and $D(4, K)$ is a radical complement of A. In particular, A is solvable. The set T of matrices of the form $\begin{pmatrix} a & 0_K & 0_K & 0_K \\ b & c & 0_K & 0_K \\ 0_K & 0_K & d & 0_K \\ 0_K & 0_K & 0_K & e \end{pmatrix}$ is now further analyzed within A. T is a K-subalgebra of A. A is solvable, and thus the nilradical of T is – based on part (i) of theorem 26 – exactly the set of matrices of the form $\begin{pmatrix} 0_K & 0_K & 0_K & 0_K \\ b & 0_K & 0_K & 0_K \\ 0_K & 0_K & 0_K & 0_K \\ 0_K & 0_K & 0_K & 0_K \end{pmatrix}$. In addition, $D(4, K)$ is a radical complement in T. Let $r := \begin{pmatrix} 0_K & 0_K & 0_K & 0_K \\ 0_K & 0_K & 0_K & 0_K \\ 0_K & 0_K & 0_K & 0_K \\ 1_K & 0_K & 0_K & 0_K \end{pmatrix}$. $r \in rad(A)$ and $r^2 = 0_A$ are valid. Based on part (v) of theorem 3 we calculate $(1_A + r)^{-1} = 1_A - r$, and the section 1.3.2 lets us calculate $D(4, K)^{1_A + r}$ is representable by matrices of the form $\begin{pmatrix} a & 0_K & 0_K & 0_K \\ 0_K & b & 0_K & 0_K \\ 0_K & 0_K & c & 0_K \\ d-a & 0_K & 0_K & d \end{pmatrix}$. $D(4, K)^{1_A + r} \cap T$ is a 3-dimensional algebra, and hence this intersection is no radical complement in T. Now we focus on the right ideal R of A generated by the matrix $\begin{pmatrix} 1_K & 0_K & 0_K & 0_K \\ 1_K & 1_K & 0_K & 0_K \\ 1_K & 1_K & 0_K & 0_K \\ 1_K & 1_K & 0_K & 0_K \end{pmatrix}$. This right ideal is representable by matrices of the form $\begin{pmatrix} a & 0_K & 0_K & 0_K \\ a+c & b & 0_K & 0_K \\ a+c & b & 0_K & 0_K \\ a+c & b & 0_K & 0_K \end{pmatrix}$. By using part (i) of theorem 26 we deduce $rad(R) = rad(A) \cap R$. Hence, $D(4, K) \cap R$ is a radical complement in R. But it is straightforward to prove that $dim_K(D(4, K)^{1_A + r} \cap R) = 1$ is valid.◇

Within the last two sections of this chapter we focus on the algebra of lower triangular matrices. We illustrate the results proven so far based on this algebra in small dimension and we analyse the importance of it for all solvable associative algebras.

3.4 A summarizing example

Let A be the algebra within example 1. We begin to determine all subalgebras of A. For calculating the one-dimensional subalgebras of A we have to determine the idempotent and the nilpotent elements of class 2 (because a one-dimensional K-algebra is either nilpotent or isomorphic to K). Let $k, l, m \in K$ and $a := k1_A + le + mr$. The condition

$$(*) \; a^2 = k^2 1_A + (2kl + l^2)e + (2km + lm)r$$

is valid. Hence, $a^2 = 0_A$ is true if and only if $a \in rad(A)$ is valid. We assume $a^2 = a$. If $k = 0_K$ is valid, then $l = 1_K$ and $a = e + mr$ are true. Based on $(*)$ the element a is indeed an idempotent of A. Let $k \neq 0_K$. In this case $k = 1_K$, $l(l+1_K) = 0_K$ and $m(1+l) = 0_K$ are valid. Within the case $l = 0_K$ we calculate $m = 0_K$ and $a = 1_A$. If $l = -1_K$ is true, then $a = 1_A - e + mr$ is valid which is – by using $(*)$ – again an idempotent of A.

Let S be a two-dimensional subalgebra of A. $S \cap rad(A)$ is zero- or one-dimensional. The first case leads to a radical complement of A. Within the second case let $\{r, k1_A + le\}$ a K-basis of S. We define $s := k1_A + le$ and calculate

$$\begin{aligned} r^2 &= 0_A, \\ rs &= kr, \\ sr &= (k+l)r \text{ and} \\ s^2 &= k^2 1_A + (2_K kl + l^2)e. \end{aligned}$$

Thus, $\langle s \rangle_K$ is a one-dimensional subalgebra of A. By the proven results we can choose $s \in \{1_A, e, 1_A - e\}$.

The following Hasse diagram illustrates the structure of A. The statements within the table attached can be proven straightforward by the results shown so far and may be done by the reader as an exercise.

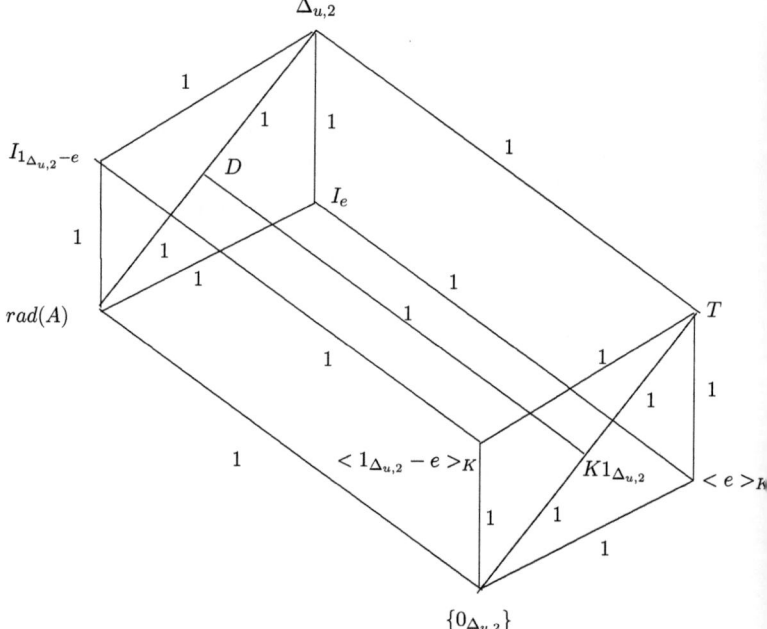

Solvable algebras

dimension	subalgebras	properties
0	$\{0_A\}$	ideal, separable, orbit under $rad(A)$ of length 1, $A^{(2)}$
1	$rad(A) = \langle r \rangle_K$	ideal, set of all nilpotent elements of A, $A^{(1)}$
	$\langle e + mr \rangle_K$	left ideal, separable, isomorphic to K, orbit under $rad(A)$ of length $\mid rad(A) \mid$
	$\langle 1_A - e + mr \rangle_K$	right ideal, separable, isomorphic to K, orbit under $rad(A)$ of length $\mid rad(A) \mid$
	$\langle 1_A \rangle_K$	separable, isomorphic to K, center of A orbit under $rad(A)$ of length 1 intersection of all radical complements of A
2	$T_m := \langle 1_A, e + mr \rangle_K$	separable, isomorphic to K^2, radical complements of A, orbit under $rad(A)$ of length $\mid rad(A) \mid$
	$D := \langle 1_A, r \rangle_K$	separable factor algebra by its nilradical, commutative, K is a splitting field, possesses exactly one radical complement, $rad(D) = rad(A) \cap D = rad(A)$, $D = rad(D) \oplus_K T_m \cap D$ for all $m \in K$
	$I_e := \langle e, r \rangle_K$	ideal, factor algebra by its nilradical, $\mid rad(A) \mid$-many radical complements which are left ideals in I_e, $rad(I_e) = rad(A) \cap I_e = rad(A)$, $I_e = rad(I_e) \oplus_K (T_m \cap I_e)$ for all $m \in K$, K is a splitting field
	$I_{1_A-e} := \langle 1_A - e, r \rangle_K$	ideal, separable factor algebra by its nilradical, $\mid rad(A) \mid$-many radical complements which are right ideals in I_{1_A-e}, $rad(I_{1_A-e}) = rad(A) \cap I_{1_A-e} = rad(A)$, $I_{1_A-e} = rad(A) \oplus_K (T_m \cap I_{1_A-e})$ for all $m \in K$, K is a splitting field
3	A	separable factor algebra by its nilradical, K is a splitting field, $st(A) = 2$ $\mid rad(A) \mid$-many radical complements, 5 orbits of separable subalgebras, every semisimple subalgebra is separable, 4 classes of isomorphism of 2-dimensional subalgebras, 2 classes of isomorphism of 1-dimensional subalgebras

Within this example every separable subalgebra of A possesses maximal or minimal orbit length. But this is not a general theorem for solvable splitting associative algebras. This may be true because of low dimension. If we focus on the algebra $A := \Delta_{u,3}$, then it is straightforward to prove that two-dimensional unitary separable subalgebra generated by unit matrix and the matrix $\begin{pmatrix} 1_K & 0_K & 0_K \\ 0_K & 0_K & 0_K \\ 0_K & 0_K & 1_K \end{pmatrix}$ the statement $C_A(T) \cap rad(A) = \langle \begin{pmatrix} 0_K & 0_K & 0_K \\ 0_K & 0_K & 0_K \\ 1_K & 0_K & 0_K \end{pmatrix} \rangle_K$ is true. Thus, this subalgebras does not possess maximal or minimal orbit length (see part (i) of remark 16).⋄

3.5 Lower triangular matrices and solvable associative algebras

Within the previous section we have illustrated the results for solvable associative algebras based on lower triangular matric algebras. Now we focus on the question whether solvable associative algebras and lower triangular matrix algebras are closely connected. This topic is motivated by a corresponding theorem for Lie algebras: the theorem of Lie is valid for solvable Lie algebras over algebraical closed fields of characteristic zero. Such kind of algebras are Lie isomorphic modulo its center to a Lie subalgebra of the Lie algebra of lower triangular matrices over that field. A similar result is valid for solvable associative algebras, too.

Proposition 7 *Let K be a field and A a finite-dimensional associative K-algebra. If every irreducible A-algebra module is one-dimensional, then A is solvable.*

Proof. The proof is using induction on $dim_K(A)$. If $dim_K(A) \leq 1$, then A is solvable. Let I be a minimal K-ideal of A. Thus, I is an irreducible A-module which is (based on our assumption) one-dimensional. In particular, I is solvable. A/I respects the assumption for our induction argument. Therefor, this factor algebra is solvable, too. The proof is finished by using part (iv) of theorem 19.⋄

Theorem 28 *(irreducible modules of solvable algebras) Let K be an algebraical closed field, $char(K) = 0$ and A a finite-dimensional associative solvable K-algebra. Every irreducible A-algebra module of A is one-dimensional.*

Proof. Based on proposition 4 the algebra A° is a finite-dimensional solvable K-Lie algebra. In addition, every irreducible A-module is an irreducible A°-module. Hence, the proof is finished by using the theorem of Lie for solvable Lie algebras.⋄

Corollary 5 *Let K be an algebraical closed field, $char(K) = 0$ and A a finite-dimensional associative solvable K-algebra. The following statements are valid:*

Solvable algebras 123

(i) *If V is a n-dimensional A-algebra module with respect to δ, then a K-basis B of V exists such that for all $a \in A$ the matrix $M_B(a\delta)$ is lower triangular in $K^{n \times n}$.*

(ii) *A K-basis B of A exists such that for all $a \in A$ the matrix $M_B(a\rho)$ is lower triangular in $K^{dim_K(A) \times dim_K(A)}$.*

Proof. ad(i): Let $\mathcal{K} := (V_0, ..., V_r)$ be a chain of A-algebra submodules of V such that $V_0 = \{0_V\}$, $V_r = V$ and for all $i \in \underline{r}$ the module V_i/V_{i-1} is an irreducible A-module. Based on theorem 28 for all $i \in \underline{r}$ the condition $dim_K(V_i/V_{i-1}) = 1$ is valid. If B is K-basis of V well adapted to the chain $\mathcal{K}$, then B fulfills part (i).

ad(ii): This statement is a direct consequence of part (i).◇

Remark 19 *Let A be an associative K-algebra, I a right ideal of A and ρ_I the right regular representation of I. The following statements are valid:*

(i) *If I is left unitary, then ρ_I is injective.*

(ii) *$ker\rho_I$ is a zero-Ideal of A.*◇

Corollary 7 *(triangulation of solvable algebras) Let K be an algebraical closed field, $char(K) = 0$ and A a finite-dimensional associative solvable K-algebra. The following statements are valid:*

(i) *If A is left unitary, then A is $\mathcal{A}$-isomorphic to a K-subalgebra of the algebra of lower triangular matrices of $K^{dim_K(A) \times dim_K(A)}$.*

(ii) *$ker\rho$ is a zero-ideal of A and $A/ker\rho$ is $\mathcal{A}$-isomorphic to a K-subalgebra of the algebra of lower triangular matrices of $K^{dim_K(A) \times dim_K(A)}$.*

Proof. The proof is a consequence of part (ii) of corollary 5 and remark 19.◇

Note 3 *Let K be an algebraical closed field, $char(K) = 0$, A a finite-dimensional associative K-algebra and $\rho_{AUF(A)}$ the right regular representation of $AUF(A)$.*
Within the algebra A the following situation of its structure is valid (see corollary 7 and remark 19):
The factor algebra $A/AUF(A)$ is $\mathcal{A}$-isomorphic to a direct sum of non-commutative matrix algebras over K.
If $AUF(A)$ left unitary, then $AUF(A)$ is $\mathcal{A}$-isomorphic to a subalgebra of the algebra of lower triangular matrices of $K^{dim_K(AUF(A)) \times dim_K(AUF(A))}$.
More general, $ker\rho_{AUF(A)}$ is a zero-ideal of A and $AUF(A)/ker\rho_{AUF(A)}$ is $\mathcal{A}$-isomorphic to subalgebra of the algebra of lower triangular matrices of

$K^{dim_K(AUF(A)) \times dim_K(AUF(A))}$.

One question is if corollary 7 is valid for other types of fields. For this, the following example is presented.◇

Example 9 Let N be a finite-dimensional associative nilpotent $\mathbb{Q}$-algebra and $A := N \oplus \mathbb{Q}(i)$. The statements $rad(A) \cong_{\mathcal{A}} N$ and $A/rad(A) \cong_{\mathcal{A}} \mathbb{Q}(i)$ are valid. In particular, $\mathbb{Q}$ is no splitting field of A. If $A/ker\rho$ would be $\mathcal{A}$-isomorphic to a subalgebra of $\mathbb{Q}^{dim_K(A) \times dim_K(A)}$, then – based on part (iii) of theorem 26 – $\mathbb{Q}$ would be a splitting field of $A/ker\rho$. But the factor algebra by the nilradical of $A/ker\rho$ is $\mathcal{A}$-isomorphic to $A/rad(A)$.◇

3.6 Open-ended questions and exercises

Open-ended question 2 *(i) Is it possible to transfer the results of this chapter to the associated Jordan algebra?*

(ii) Analyze more Lie properties of the associated Lie algebra based on an associative algebra (e.g. minimal non-solvable algebras and supersolvable algebras).

(iii) Does a close connection between the classes of solvability and the derived series for an associative algebra, its associated Lie-algebra and its group of units exist?

(iv) Solve the previous topic for group algebras.

(v) Determine the radical of $<,>_{\lambda,\rho}$.

(vi) On what terms is $<,>_{\lambda,\rho}$ associative?

(vii) Is theorem 24 also valid for $char(K) \neq 0$?

(viii) Determine the structure of $A/auf(A)$ based on the structure of A. On what terms is the condition $st(AUF(A)) = st(A/auf(A))$ valid?

Excercise 192 Use the article of M. R. Bremner (see [4]) and theorem 2 to derive a method for calculating the nilradical of a finite-dimensional associative K algebra in the case $char(K) = 0$. Apply this calculation to the algebra of lower triangular matrices over K in dimension ≤ 3.

Excercise 193 Try to describe in what way the nilpotency index of a nilpotent associative algebra is connected to chains of ideals of the algebra possessing successive zero factor algebras. What is the importance of the chain of powers of the nilradical?

Excercise 194 Draw a Hasse diagram illustrating the content of remark 9.

Solvable algebras

Excercise 195 Draw a Hasse diagram illustrating the content of remark 3.

Excercise 196 Let A be an associative solvable K-algebra. For all $n \in \mathbb{N}$ the statement $Q(A)^{(n)} \subseteq A^{(n)}$ is valid. In particular, $Q(A)$ is solvable and $st(Q(A)) \leq st(A)$ is valid.

Excercise 197 Let A be the algebra of upper or lower triangular matrices over an arbitrary field and $r, s \in \mathbb{N}$. Prove the identity $rad(A)^{<r>} \circ rad(A)^{<s>} = rad(A)^{<r+s>}$. Which general statement of this chapter is enhanced by this identity?

Excercise 198 *(eAe)* Let K be a field, A a solvable finite-dimensional associative unitary K-algebra possessing a separable factor algebra by its nilradical, T a radical complement of $rad(A)$ in A and e an idempotent of A. Analyze the following topics for the algebra eAe (see exercise 173):

(i) $A = rad(A) \oplus T$ is valid, and thus $eAe = e\,rad(A)\,e \oplus eTe$ is true. Prove that eTe is a separable radical complement. Use that every subalgebra of a commutative separable algebra is separable and commutative. Hence, eAe is a solvable finite-dimensional associative unitary K-algebra possessing a separable factor algebra by its nilradical. Determine its unit element.

(ii) True or false: eAe is solvable if and only if A is solvable.

(iii) Is it possible to connect or bound the derived series and classes of solvability of eAe, of $(eAe)°$ and of $E(eAe)$ with the ones of A, $A°$ and $E(A)$? If the exercise is to complex, then assume that e is a central idempotent.

Excercise 199 *(zero-extension)* Let A be a K-algebra based on a multiplication $\cdot$. On $B := A \times A$ we define a new multiplication $\odot$ by

$$(a, x) \odot (b, y) := (ab, ay + xb)$$

(see exercise 174). Analyze the following topics:

(i) If A is commutative and semisimple, then B is solvable.

(ii) On what terms is B solvable?

(iii) On what terms is $B°$ solvable?

(iv) On what terms is $E(B)$ solvable?

(v) If B is solvable, then determine the classes of solvability for B, $B°$ and $E(B)$. Illustrate the connections to the ones of A.

If the exercise is to complex, then assume that A is semisimple or separable.

Excercise 200 Let A, B be associative K-algebras, T a subalgebra and I an ideal of A. Analyze the following topics:

(i) What are the connections between the derived series and classes of solvability of A, B and of $A \times B$?

(ii) Is it possible to bound the class of solvability of T by the class of solvability of A?

(iii) Is it possible to bound the class of solvability of A/I by the classes of solvabilities of A and I?

(iv) Let A be solvable and $rad(A)$ nilpotent. Is it possible to bound the class of solvability of A by the class of nilpotency of $rad(A)$?

Excercise 201 Let A be an associative K-algebra. Prove that $rad(A) \subseteq Nil(A)$ is valid. Find an example of an algebra such that $rad(A)$ is not equal to $Nil(A)$. What is the consequence within this example for the structure of the factor algebra by the nilradical?

Excercise 202 Let A be a nilpotent associative K-algebra. Prove that A° is nilpotent and that $cl(A^\circ) \leq cl(A)$ is valid.

Excercise 203 Let A be an associative K-algebra and T be a central subalgebra of A. Prove that A is a T-algebra. Assume that T is a field. What are the connections between the dimensions of A over K and of A over T? Do prominent examples of algebras A exist such that T is automatically a field?

Excercise 204 Prove that the associated Lie algebra of an associative algebra is a Lie algebra.

Excercise 205 Formulate the theorem of Lie for solvable Lie algebras. What is its relevance within this chapter?

Excercise 206 Define the Killing form for Lie algebras and state its relevance for solvable Lie algebras. What is its relevance within this chapter?

Excercise 207 Analyze within exercise 203 if A° is a T-Lie algebra.

Excercise 208 Let us focus on the nth powers of an associative K-algebra. Is this set a subspace, a subalgebra, a right ideal, a left ideal or an ideal?

Excercise 209 Prove for an associative K-algebra A that its opposite algebra A^{op} is solvable if and only if A is solvable. Analyze the connection between their nilradicals and derived series. What is the consequence for their classes of solvability?

Solvable algebras   127

Excercise 210 *Let A be an associative K-algebra. What are the connections between the standard bilinear forms and $<,>_{\lambda,\rho}$ on A and A^{op}?*

Excercise 211 *Let K be a field and A a finite-dimensional associative unitary K-algebra. Prove that $AUF(A)/rad(A)$ possesses an ideal complement in $A/rad(A)$. In which way is it possible to describe this complement? Is it necessary for the existence of this ideal complement that A is unitary? Is this ideal complement unique? Calculate this decomposition in the case of $\Delta_{u,3} \times \mathbb{H}$ and $\Delta_{u,3} \times \mathbb{H} \times \mathbb{H}$ (regarded as $\mathbb{R}$-algebras). In addition, determine the dimension of the ideals of these decompositions in $A/rad(A)$.*

Excercise 212 *Let K be a field, $n \in \mathbb{N}$, $18 \leq n \leq 35$, $A := \Delta_{u,n}$ and $B := \Delta_{o,n}$. Determine and illustrate within a table the classes for solvability for $A, B, A^\circ, B^\circ, E(A)$ and $E(B)$. What is the relevance of K within this calculations?*

Excercise 213 *Let A be a finite-dimensional associative K-algebra. Which of the following statements are true in general and which are equivalent to A being solvable:*

(i) $AUF(A) = 0$

(ii) $auf(A) = 0$

(iii) $AUF(A) = A$

(iv) $auf(A) = A$

(v) $auf(A) \leq AUF(A)$

(vi) $AUF(A) \leq auf(A)$

(vii) $rad(A) \leq AUF(A)$

(viii) $Nil(A) \leq AUF(A)$

(ix) $rad(A) \leq auf(A)$

(x) $Nil(A) \leq auf(A)$

(xi) $auf(A) \leq rad(A)$

(xii) $auf(A) \leq Nil(A)$

(xiii) $AUF(A) \leq rad(A)$

(xiv) $AUF(A) \leq Nil(A)$?

Excercise 214 *Let A be a finite-dimensional associative unitary K-algebra with associated Lie algebra A° and group of units $E(A)$. True or false:*

(i) If A is solvable, then A° is solvable.

(ii) If A° is solvable, then A is solvable.

(iii) If A° is solvable and $char(K) \neq 2$ is valid, then A is solvable.

(iv) If A is solvable, then $E(A)$ is solvable.

(v) If $E(A)$ is solvable, then A is solvable.

(vi) If $E(A)$ is solvable and $char(K) \neq 2$ is valid, then A is solvable.

(vii) If $E(A)$ is solvable and $char(K) \neq 2$ as well as $|K| \geq 5$ are valid, then A is solvable.

(viii) If $E(A)$ is solvable and $char(K) = 0$ is valid, then A is solvable.

(ix) If A° is solvable, then $E(A)$ is solvable.

(x) If A° is solvable and $char(K) \neq 2$ is valid, then $E(A)$ is solvable.

(xi) If $E(A)$ is solvable, then A° is solvable.

(xii) If $E(A)$ is solvable and $char(K) \neq 2$ is valid, then A° is solvable.

(xiii) If $E(A)$ is solvable and $char(K) \neq 2$ as well as $|K| \geq 5$ is valid, then A° is solvable.

(Tip: see also the appendix chapter of this volume)

Excercise 215 *Analyze exercise 214 for $A := KG$ (K a field and G a finite group) and for the algebra of lower triangular matrices (in finite dimension) over a field.*

Excercise 216 *Let A be an associative K-algebra, A° its associated Lie-algebra and T a central subalgebra of A. Prove that A and A° are T-algebras. True or false:*

(i) If A is solvable as T-algebra, then A is solvable as K-algebra.

(ii) If A is solvable as K-algebra, then A is solvable as T-algebra.

(iii) If A° is solvable as T-algebra, then A is solvable as K-algebra.

(iv) If A° is solvable as K-algebra, then A is solvable as T-algebra.

(v) If A is solvable as T-algebra, then A° is solvable as T-algebra.

(vi) If A° is solvable as T-algebra, then A is solvable as T-algebra.

If one of the statements is false, then, in addition, analyze on what terms it is true.

Solvable algebras

Excercise 217 Solve exercise 216 by replacing the word solvable by nilpotent.

Excercise 218 Prove remark 17 in details.

Excercise 219 Within the construction 1 prove part (v) in details.

Excercise 220 Prove the induction argument within example 15 in details.

Excercise 221 Prove example 16 in details.

Excercise 222 Prove proposition 4 in details.

Excercise 223 Prove that radicals of associative bilinear forms on an associative algebras are ideals. Is it necessary to assume that the algebra is associative?

Excercise 224 Let D be a central associative semisimple right artian K-algebra. If D is solvable, then $D = K \cdot 1_D$ is valid.

Excercise 225 Let K be a field, D be a K-quaternion algebra and $char(K) = 2$. Analyze the following statements:

(i) Is D nilpotent? Determine the associative powers of D.

(ii) Is D° nilpotent? Determine the descending Lie-central series of D.

(iii) Is D solvable? Determine the derived series of D.

(iv) Is D° solvable? Determine the derived series of D°.

Excercise 226 Let K be a field, D be a K-quaternion algebra and $char(K) \neq 2$. Analyze the following statements:

(i) Is D nilpotent? Determine the associative powers of D.

(ii) Is D° nilpotent? Determine the descending Lie-central series of D.

(iii) Is D solvable? Determine the derived series of D.

(iv) Is D° solvable? Determine the derived series of D°.

Excercise 227 Illustrate how the structure of a finite-dimensional associative algebra based on an algebraical closed field can be described by using zero algebras, lower triangular matrices, solvable algebras and non-commutative matrix algebras.

Excercise 228 Let A be an associative finite-dimensional K-algebra. Prove the statement $rad(A) \leq Nil(A) \leq rad(<,>_{\lambda,\rho})$.

Excercise 229 *Every associative homomorphism between two associative algebras is a Lie-homomorphism between their associated Lie-algebras. What is the consequence for algebra representations?*

Excercise 230 *Let G be a finite group and K be a field with $char(K) = 0$. By using the bilinear form $<,>_{\lambda,\rho}$ analyze on what terms KG is solvable. Which result is to be expected?*

Excercise 231 *Determine $Nil(A)$ for the following algebras A:*

(i) $\mathbb{C}^{2\times 2}$

(ii) $\mathbb{H}$

(iii) $\mathbb{Q}$

(iv) a division algebra

(v) a direct product of algebras

(vi) a local algebra

(vii) a solvable algebra

(viii) a basic algebra

(ix) $\Delta_{u,n}$

(x) $\Delta_{o,n}$

(xi) a quaternion algebra

(xii) $K^{n\times n}$, K a field, $n \in \mathbb{N}$

(xiii) eAe, e a (central) idempotent (see exercise 173)

(xiv) a zero extension of a (separable) algebra (see exercise 174)

(xv) a factor algebra with respect to a nil ideal

(xvi) a factor algebra with respect to a nilpotent ideal

(xvii) a factor algebra with respect to an arbitrary ideal

(xviii) the factor algebra of A modulo $\langle Nil(A)\rangle_{\lhd A}$.

Decide within each example whether $Nil(A)$ and $rad(A)$ are identical. If they are not identical, then analyze on what terms they are identical. Is there are a general result existing clarifying the conditions for $Nil(A)$ and $rad(A)$ being identical?

Solvable algebras

Excercise 232 Let $n \in \mathbb{N}$ and K be a field. Apply the bilinear form $<,>_{\lambda,\rho}$ within $K^{n \times n}$ on all subspaces linked to one column resp. one row. Is the form associative on these spaces? Determine the radical of this form. (A subspace linked to one column resp. line is the set of all matrices for which only within this column resp. line entries different from zero exist.)

Excercise 233 Is it possible to enhance exercise 232 in a way such that instead of the field K an associative (and solvable) K-algebra A is used?

Excercise 234 Prove all statements within example 3.4.

Excercise 235 Let K be a field and a an element of a finite-dimensional unitary associative K-algebra A. Analyze in what way the following statements are linked:

(i) a is nilpotent.

(ii) a is contained in the nilradical of A.

(iii) The trace of $a\rho$ is exactly 0.

(iv) $a\rho$ is nilpotent.

(v) $a\lambda$ is nilpotent.

(vi) The trace of $a\lambda$ is exactly 0.

(vii) $a(\lambda + \rho)$ is nilpotent.

(viii) The trace of $a(\lambda + \rho)$ is exactly 0.

(ix) $a(\lambda - \rho)$ is nilpotent.

(x) The trace of $a(\lambda - \rho)$ is exactly 0.

(xi) $a(\lambda\rho)$ is nilpotent.

(xii) The trace of $a(\lambda\rho)$ is exactly 0.

Excercise 236 Let $A := \Delta_{u,3}$ regarded as $\mathbb{Q}$-algebra. We focus on the following subspaces: $T = D(n,3)$-diagonal matrices, Z_3-line-subspace linked to the third line, S_1-column subspace linked to the first column, $Z = Z(A)$, $I = rad(A)^2$ and $M = rad(A)^2 \oplus C_T(rad(A)^2)$. Determine the nilradical of and one radical complement for Z_3, S_1, Z, I, M, A/I and $Z_3 \times S_1$. What are the corresponding dimensions? Does a radical complement of A exist which can be used to derive a radical complement for all mentioned algebras? Represent M in terms of matrices. Which of the structures Z_3, S_1, Z, I, M, A/I and $Z_3 \times S_1$ are Lie-nilpotent or Lie-solvable?

Excercise 237 Let K be a field, $n \in \mathbb{N}$, p a prime number and G a finite group. Within the following examples analyze whether KG, $(KG)^\circ$ or $E(KG)$ is solvable:

(i) G arbitrary, $K = \mathbb{C}$

(ii) G arbitrary, $K = \mathbb{Q}$

(iii) G arbitrary, $K = \mathbb{R}$

(iv) G arbitrary, $char(K) = 0$

(v) G commutative

(vi) G is a p-group and $K = GF(p)$

(vii) G is nilpotent and $K = GF(p)$

(viii) G is solvable and $K = GF(p)$

(ix) $G = A_n$ and $K = GF(p)$

(x) $G = S_n$ and $K = GF(p)$

(xi) $G = Q_8$ and $K = GF(p)$

(xii) G non-Abelian and hamiltonian and $K = GF(p)$.

In addition, determine a basis and the dimension of $Aug(KG)$ for all examples!

Excercise 238 Let $A := \Delta_{o,4}$ as an algebra over $\mathbb{C}$. Apply theorem 24 to this algebra. Is the bilinear form mentioned there symmetrically? Calculate its radical! Calculate the bilinear form on its basis. Is A solvable or Lie solvable?

Excercise 239 Solve exercise 238 for the fields $GF(2)$ and $GF(5)$.

Excercise 240 Do a research in the literature for results related to the determination of the class of solvability of a Lie solvable group algebra and of a solvable group of units. Compare the results.

Excercise 241 Let K be a field, p a prime number, $char(K) = p$ and G a finite p-group. Prove that $Aug(KG)$ is a nilpotent ideal of KG. In particular, $K1_G$ is radical complement of $rad(KG) = Aug(KG)$ in KG. Hence, the associative derived series of K and its nilradical $Aug(KG)$ are closely connected. These can be bounded by the 2^n-th associative powers of the nilradical, and we conclude $st(KG^\circ) \leq \lfloor log_2(cl(rad(KG))) \rfloor$. Prove all of these statements in details.

Solvable algebras

Excercise 242 Let K be a field, G a finite group and N a normal subgroup of G. Prove that $KGAug(KN) = Aug(KN)KG$ is an ideal of KG and that its factor algebra is isomorphic to $K(G/N)$. Determine a K-linear basis of this ideal. For the groups Q_8, D_8, S_3 and A_4 calculate this basis in details. Within these examples determine all normal subgroups N and fields K such that $KGAug(KN)$ or $Aug(KN)$ is nilpotent. Analyze in general if both structures $KGAug(KN)$ and $Aug(KN)$ are nilpotent or not. Is it possible to reduce the calculation of the class of nilpotency of $KGAug(KN)$ to the one of $Aug(KN)$? How is this reduction usable for a p-Sylow subgroup? In what way is exercise 241 relevant for this topic?

Excercise 243 Let p be a prime number, K a field, $char(K) = p$, G a finite p-group and $z \in G$ such that $G = \langle z \rangle_{\mathcal{G}}$ is valid. Prove the statement $rad(KG) = (z-1_G)KG$. In addition, find connections between $cl(rad(KG))$, $cl(1-z)$ and $o(z)$. (Tip: exercise 241)

Excercise 244 Let p be a prime number, K a field, $char(K) = p$ and G a finite p-group possessing a cyclic derived subgroup. Use exercises 241, 242 and 243 to calculate an upper bound for the class of solvability for $(KG)^\circ$. Within the case $p = 2$ calculate this bound for a quaternion group, a dihedral and a semi-dihedral group of prime power order 2.

Excercise 245 Let A be a finite-dimensional associative K-algebra. $auf(A)$ is the smallest ideal such that $A/auf(A)$ is solvable. We define recursively $auf(A)^n := auf(auf(A)^{n-1})$ for all $n \geq 2$. Prove or disprove: The following statements are equivalent:

(i) A is solvable.

(ii) $auf(A) = 0$

(iii) An element $n \in \mathbb{N}$ exists such that $auf(A)^n = 0$ is valid.

In addition, prove that the series $(auf(A)^n)_{n \in \mathbb{N}}$ stabilizes after finite many steps. Calculate this series for $\Delta_{u,n}$ and for $\Delta_{u,n} \times \mathbb{H}$ (both regarded as $\mathbb{R}$-algebras).

Excercise 246 Formulate and prove similar results for $AUF(A)$ instead of $auf(A)$ based on exercise 245.

Excercise 247 Let K be field of positive characteristic p, $n \in \mathbb{N}$, A an associative K-algebra and $a \in A$. $a\lambda$ and $a\rho$ commute with each other. Deduce a calculation method for $(a\lambda + a\rho)^{p^n}$! Let $A := \Delta_{u,2}$. For which elements $a \in A$ is the endomorphism $a(\lambda + \rho)$ nilpotent? Is it possible to represent this endomorphism by using the basis of A?

Excercise 248 Let A be an associative K-algebra. $A^{<n>}$ is an ideal of A for all $n \in \mathbb{N}$, and for all $n \leq m$ the identity $A^{<n>} \geq A^{<m>}$ is valid. If A is unitary, then all of these ideals are identical with A. If A is right artian, then this chain of ideals stabilizes after finite many steps. Let $A^{<n>}$ be the smallest ideal within this chain. Prove or disprove the identity $A = rad(A) + A^{<n>}$. A is nilpotent if and only if the chain stabilizes at 0.

Excercise 249 Which of the following associative algebras are solvable and/or unitary?

 (i) a zero-algebra

 (ii) a commutative algebra

 (iii) a nilpotent algebra

 (iv) an arbitrary group algebra

 (v) a basic algebra

 (vi) a local algebra

 (vii) a field

(viii) a division algebra

 (ix) direct product of solvable algebras

 (x) tensor product of solvable algebras

 (xi) the algebra of lower triangular matrices over an arbitrary field

 (xii) the algebra of upper triangular matrices over an arbitrary field

(xiii) subalgebras of solvable algebras

(xiv) factor algebras of solvable algebras

 (xv) full matrix algebras over fields

(xvi) full matrix algebras over solvable algebras

(xvii) $GF(3)S_3$

(xviii) $GF(3)A_3$

 (xix) $GF(2)S_5$

 (xx) $\mathbb{Q}A_7$

 (xxi) $GF(p)P$ such that P is a p-group

Solvable algebras

(xxii) $GF(p)(P \times Q)$ such that P is a p-group and Q is an Abelian group.

If one of the following algebras is not solvable or not unitary, then analyze on what terms this condition is valid.

Excercise 250 Analyze whether the parts (iv) and (v) within theorem 23 are equivalent to the following statements:

(i) For all $a \in A \circ A$ the statement $<a, a>_{\lambda,\rho} = 0 = <a^2, 1_A>_{\lambda,\rho}$ is valid.

(ii) For all $a, b, c \in A$ the identity $<a \circ b, c>_{\lambda,\rho} = 0 = <(a \circ b)c, 1_A>_{\lambda,\rho}$ is valid.

Excercise 251 Under the conditions of theorem 23 verify: If $<,>_{\lambda,\rho}$ is associative, then its radical is an ideal. If a is an element of this radical, then $a(\lambda + \rho)$ is nilpotent. (Hint: exercise 252)

Excercise 252 This exercise is used within the proof of exercise 42 on page 277 in [37]: Let $n \in \mathbb{N}$, V a n-dimensional K-space and $f \in End_K(V)$. Let $char_{f,K} = t^n + \sum_{i=0}^{n-1} c_i t^i$ be the characteristical polynomial of f over K. Prove the following identities:

$$\begin{aligned} 0 &= tr(f) + c_1 \\ 0 &= tr(f^2) + c_1 tr(f) + 2c_2 \\ &\cdots \\ 0 &= tr(f^n) + c_1 tr(f^{n-1}) + \cdots + c_{n-1} tr(f) + nc_n. \end{aligned}$$

On what terms is f nilpotent? What is the relevance of this exercise within this chapter? Analyze the usage of this exercise within this chapter: prove the linked result not only for $char(K) = 0$ but also for the more general case $(char(K), dim_K(A)) = 1$. (Hint: standard matrix trace bilinear form)

Excercise 253 Prove the following theorem of Wedderburn: Let A be a finite-dimensional associative K-algebra and I an ideal of A which is K-linear generated by nilpotents elements. I is nilpotent. (Tip: A right artian associative algebra is nilpotent if every element is nilpotent. (These algebras are called nil).)

Excercise 254 Let A, B be associative K-algebras. True or false: $(A \otimes B)°$ and $A° \otimes B°$ are isomorphic.

Excercise 255 Let A, B be right artian associative K-algebras. Prove that $rad(A) \otimes B + A \otimes rad(B)$ is a nilpotent ideal of $A \otimes B$.

Excercise 256 Let A be an associative K-algebra and I an ideal of A. True or false: $(A/I)°$ is identical to $A°/I°$.

Excercise 257 Let A be an associative K-algebra and I, J ideals of A. True or false: If I, J are nilpotent (solvable), then $I + J$ is nilpotent (solvable).

Excercise 258 Let A be an associative K-algebra and I, J ideals of A. True or false: If $A/I, A/J$ are nilpotent (solvable), then $A/(I + J)$ is nilpotent (solvable).

Excercise 259 We focus on counterexample 4. Calculate the counterexample again by using the opposite algebra. What are the results?

Excercise 260 Let A be an associative K-algebra and $a, b \in A$. Prove $Nil(A) \cdot C_A(Nil(A)) \subseteq Nil(A)$.

Excercise 261 Within the calculation 1 prove the key idea in details which is linking the sum and the product.

Chapter 4

Generalized quaternion algebras

4.1 Definition and isomorphism

Definition and remark 9 *(generalized quaternion algebras)* Let K be a field, $a, b \in K$ and $A(a, b, K)$ the 4-dimensional unitary K-algebra possessing the K-basis $\{1_{A(a,b,K)}, i, j, k\}$ equipped with the multiplication matrix

·	$1_{A(a,b,K)}$	i	j	k
$1_{A(a,b,K)}$	$1_{A(a,b,K)}$	i	j	k
i	i	$a1_{A(a,b,K)}$	k	aj
j	j	-k	$b1_{A(a,b,K)}$	-bi
k	k	-aj	bi	$-ab1_{A(a,b,K)}$.

The K-algebra $A(a, b, K)$ is called generalized quaternion algebra. We mention that in the literature $A(a, b, K)$ is called like this only within the case $char(K) \neq 2$ and $a, b \neq 0_K$ (see e.g. chapter 1 in [35]). If the link to the field K is obvious, then we write $A(a, b)$ instead of $A(a, b, K)$. In particular, $\mathbb{H} = A(-1, -1, \mathbb{R})$ is valid. We remark that $A(a, b)$ is associative and only in characteristic 2 commutative. Within the exercises we focus on generalized quaternion algebras in characteristic 2 as they are known in the literature.

Within this chapter we analyze the structure of the algebras $A(a, b)$. In particular, we want to verify the assumptions and results based on theorem 8. In addition, we classify these algebras up to isomorphism. For this, the following theorem is proven.

Theorem 29 *(Isomorphism theorems)* Let K be a field and $a, b \in K$. The following statements are valid:

(i) $A(a, b) \cong_{A_1} A(b, a)$

(ii) If $a \neq 0_K$ is valid, then $A(a,a) \cong_{A_1} A(a, -1_K)$ is true.

(iii) For all $c \in K \setminus \{0_K\}$ the statement $A(a,b) \cong_{A_1} A(a, c^2 b)$ is valid.

(iv) Let $char(K) = 2$. For all $c \in K$ the statement $A(a,b) \cong_{A_1} A(a, c^2 + b)$ is true.

Proof. ad(i): Let $A := A(b,a)$. $B := \{1_A, -j, i, k\}$ is a K-basis of A. The multiplication of A within this K-basis B is represented by the following matrix:

$\cdot$	-j	i	k
-j	$a1_A$	k	ai
i	-k	$b1_A$	bj=-b(-j)
k	-ai	-bj=b(-j)	$-ab1_A$.

This matrix is used to prove part (i).

ad(ii): Let $A := A(a, -1_K)$. Because of $a \neq 0_K$ the set $B := \{1_A, i, k, aj\}$ is a K-basis of A. The multiplication of A within this K-basis B is represented by the following matrix:

$\cdot$	i	k	aj
i	$a1_A$	aj	ak
k	$-aj$	$a1_A$	$-ai$
aj	$-ak$	ai	$-a^2 1_A$.

This matrix is used to prove part (ii).

ad(iii): Let $A := A(a, c^2 b)$. Because of $c \neq 0_K$ the set $B := \{1_A, i, c^{-1}j, c^{-1}k\}$ is a K-basis of A. The multiplication of A within this K-basis B is represented by the following matrix:

$\cdot$	i	$c^{-1}j$	$c^{-1}k$
i	$a1_A$	$c^{-1}k$	$ac^{-1}j$
$c^{-1}j$	$-c^{-1}k$	$b1_A$	$-bi$
$c^{-1}k$	$-ac^{-1}i$	bi	$-ab1_A$.

This matrix is used to prove part (iii).

ad(iv): Let $A := A(a, c^2 + b)$. The set $B := \{1_A, i, c1_A + j, ci + k\}$ is a K-basis of A. We calculate the multiplication matrix of A with respect to the K-basis B:

$$\begin{aligned}
i^2 &= a1_A \\
i(c1_A + j) &= ci + kl, \\
i(ci + k) = i^2(c1_A + k) &= a(c1_A + k), \\
(c1_A + j)^2 = c^2 1_A + j^2 &= b1_A, \\
(c1_A + j)(ci + k) = c^2 i + ck + ck + (c^2 + b)i &= bi \text{ and} \\
(ci + k)^2 = c^2 a1_A + a(c^2 + b)1_A &= ab1_A
\end{aligned}$$

are valid. This is used to prove part (iv).◇

Remark 20 *(remark to the isomorphism theorems)* Let K be a field.

(i) If $char(K) \neq 2$ is valid, then part (ii) in theorem 29 is not true for $a = 0_K$. We will prove soon that $A(0_K, 0_K)$ possesses a three-dimensional but $A(0_K, -1_K)$ a two-dimensional nilradical.
But in the case $char(K) = 2$ we use part (iv) of theorem 29 to deduce that $A(0_K, 0_K) \cong_{\mathcal{A}_1} A(0_K, 1_K)$ is valid.

(ii) If within part (iii) of theorem 29 the condition $c = 0_K$ is valid, then the statement is not correct: for $b \notin QA(K)$ we will prove that $A(0_K, b)$ is not $\mathcal{A}_1$-isomorphic to $A(0_K, 0_K)$ ist.

(iii) Part (iv) of theorem 29 is wrong in the case $char(K) \neq 2$ (see example in part (i)).◇

4.2 The case of a big nilradical

Remark 21 Let K be a field. We focus on the K-algebra $A := A(0_K, 0_K)$ and define $I := \langle i, j, k \rangle_K$. I is a three-dimensional K-ideal of A generated by nilpotent elements. The lemma in chapter 4.6 of [35] lets us deduce (by using a theorem of Wedderburn) that I is a nilpotent ideal of A possessing the codimension 1. Hence, $rad(A) = I$ is valid and $K1_A$ is a radical complement in A. In particular, $A/rad(A)$ is separable and $A \cong_{\mathcal{A}_1} (K, rad(A))$ is true. $K1_A$ is central, and thus based on part (vi) of theorem 16 the subalgebra $K1_A$ is the unique radical complement in A.
Finally, we determine the classes of solvability of A, A° and $E(A)$. $A \circ A = \langle k \rangle_K = rad(A)^2$ is true, and thus based on proposition 6 we deduce $st(A) = st(A^\circ) = 2$. $E(A)$ is not Abelian, and we conclude by using proposition 6 that $st(E(A)) = 2$ is valid.◇

4.3 The case of characteristic not two

Within this section let K be a field and $char(K) \neq 2$.

4.3.1 The case within the literature

Remark 22 Let $a, b \in K$ such that $a \neq 0_K \neq b$ is valid. We use the lemma within chapter 1.6 of [35] to deduce that $A(a, b)$ is a central-simple associative K-algebra. In particular, $A(a, b)$ is separable based on chapter 1. $A(a, b)$ is 4-dimensional, and thus $A(a, b)$ is a division algebra or $\mathcal{A}_1$-isomorphic to $K^{2 \times 2}$. The proposition in chapter 1.6 of [35] clarify the questions on what terms $A(a, b)$ is a skew field. In addition, the lemma and the proposition in chapter 1.7 of [35] shows us on what terms two quaternion algebras of this type are $\mathcal{A}_1$-isomorphic. The reader may read these results in [35] to get a deeper insight in this topic.◇

4.3.2 One component is zero

Because of part (i) of theorem 29 we only need to analyze the case $A(a, 0_K)$ such that $a \neq 0_K$ is valid.

Theorem 30 *(characterization of isomorphism)* Let $a, b \in K$ such that $a \neq 0_K \neq b$ is valid. The following statements are equivalent:

(i) $A(a, 0_K) \cong_{\mathcal{A}_1} A(b, 0_K)$

(ii) An element $l \in K$ exists such that $a = l^2 b$ and $l \neq 0_K$ are valid.

(iii) $\langle a \rangle_{QA(K)} = \langle b \rangle_{QA(K)}$

Proof. It is straightforward to prove that the parts (ii) and (iii) are equivalent. The implication from (ii) to (i) is deductable by using part (iii) of theorem 29.
Let α be a $\mathcal{A}_1$-isomorphism between $A(a, 0_K)$ and $A(b, 0_K)$ and $f_1, ..., f_4 \in K$ such that $i\alpha = f_1 1_{A(b, 0_K)} + f_2 i + f_3 j + f_4 k$. We calculate

$$a1 = i^2 \alpha = (i\alpha)^2 = (f_1^2 + f_2^2)b1 + 2f_1 f_2 i + (2f_1 f_3 + 2f_2 f_4)j + 2(f_1 f_4 + f_2 f_4 b)k.$$

By evaluating the coefficients of i we deduce $f_1 = 0_K$ or $f_2 = 0_K$. This is used to prove the theorem because of $a1 = (f_1^2 + f_2^2)b1$.◇

Theorem 30 and part (iii) of theorem 29 lets us deduce that we have to determine the structure of the algebras $A(1_K, 0_K)$ and $A(a, 0_K)$ such that $a \notin QA(K)$ is valid. For example, $t \notin QA(\mathbb{Q}(t))$ is true.

Remark 23 *(structure of the algebras $A(a,0)$)* Let $a \neq 0_K$. We define $I := \langle j, k \rangle_K$ and $T := \langle 1_{A(a, 0_K)}, i \rangle_K$. The multiplication matrix of $A := A(a, 0_K)$ lets us deduce that I is a nil, an thus also a nilpotent ideal of A and T is a subalgebra of A. $T = K[i]$ and $min_{i,K} = t^2 - a$ are valid. The minimal polynomial possesses the zeroes i and $-i$ in T. If $a \notin QA(K)$, then $min_{i,K}$ is irreducible over K and T is a field. In this case T is $\mathcal{A}_1$-isomorphic to a splitting field of the separable polynomial $min_{i,K}$ over K. In the other case $T \cong_{A_1} K^2$ is valid. Based on part (iv) of lemma 1 the algebra T is separable. Hence, it is an algebra complement of $rad(A) = I$ in A. In, particular, $A/rad(A)$ is separable and commutative.

We finalize this section by determining the classes of solvability for A, A° and $E(A)$. We calculate (which may be done by the reader as an exercise) $A \circ A = rad(A)$. Hence, (see proposition 6) $st(A) = st(A^\circ) = 2$ is valid. $E(A)$ is not Abelian, and thus proposition 6 is used to deduce $st(E(A)) = 2$.⋄

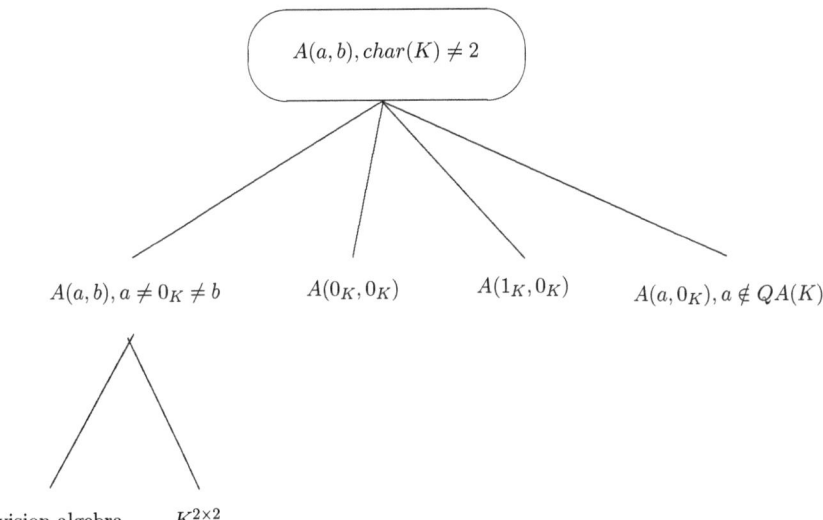

4.4 The case of characteristic equal to 2

Within this section let K be a field, $char(K) = 2$ and $a, b \in K$. The following reduction theorems for the isomorphism question covers three cases related to the elements a and b. We will show that the corresponding algebras $A(a,b)$ within these three cases are pairwise non-$\mathcal{A}_1$-isomorphic.

Theorem 31 *(reduction theorem) The following cases are valid:*

(i) $A(a,b) \cong_{A_1} A(0_K, 0_K)$

(ii) An element $a \in K \setminus QA(K)$ exists such that $A(a,b) \cong_{A_1} A(a, 0_K)$ is valid.

(iii) $a, b, ab \in K \setminus QA(K)$ is true and no pair $(g; h) \in K \times K$ exists such that $g^2 a + h^2 b = 1_K$ is valid.

Proof. Let $a, b \in QA(K)$. Based on parts (i) and (iv) of theorem 29 the statement

$$A(a,b) \cong_{A_1} A(a+a, b+b) = A(0_K, 0_K)$$

is true. Let $a \in QA(K)$ and $b \in K \setminus QA(K)$. $A(a,b) \cong_{A_1} A(b, 0_K)$ is valid by using the parts (i) and (iv) of theorem 29. Hence, case (ii) is deduced. The case $a \in K \setminus QA(K)$ and $b \in QA(K)$ can be reduce based on part (i) of theorem 29 to the previous one. Let $a, b \in K \setminus QA(K)$ and $ab \in QA(K)$. In this case $a, b, ab \neq 0_K$ is valid. We use part (iii) of theorem 29 to prove

$$A(a,b) = A(a, a^{-1}ab) \cong_{A_1} A(a, a^{-1}) \cong_{A_1} A(a, a^2 a^{-1}).$$

In addition, theorem 29 lets us deduce that

$$A(a, a^2 a^{-1}) \cong_{A_1} A(a, a) \cong_{A_1} A(a, 1_K) \cong_{A_1} A(a, 0_K)$$

is valid. Again, we have deduced case (ii). Let $a, b, ab \in K \setminus QA(K)$, and a pair $(g; h) \in K \times K$ is existing such that $g^2 a + h^2 b = 1_K$ is valid. We deduce $g \neq 0_K \neq h$. Based on theorem 29 we prove

$$A(a,b) \cong_{A_1} A(a, h^2 b) = A(a, 1_K + g^2 a) \cong_{A_1} A(a, g^2 a).$$

Furthermore, theorem 29 is used to deduce

$$A(a, g^2 a) \cong_{A_1} A(a, a) \cong_{A_1} A(a, 1_K) \cong_{A_1} A(a, 0_K).$$

Again, case (ii) is valid and the theorem is proven.⋄

All cases are valid which is the content of the next remark.

Remark 24 (i) The K-algebra $A(0_K, 0_K)$ is existing for every field K.

(ii) Let $K := GF(2)(t)$. $t \notin QA(K)$ is valid. This example is linked to case (ii) of theorem 31.

(iii) Let $K := GF(2)(t_1, t_2)$ and $F := GF(2)(t_1)$. $t_1 t_2, t_2 \notin QA(K)$ is valid, and based on $QA(K) \cap F = QA(F)$ we deduce $t_1 \notin K^2$. Let us assume a pair $(g; h) \in K \times K$ would exists such that $g^2 t_1 + h^2 t_2 = 1_K$ is true. We would deduce that elements $g, h, l \in F[t_2]$ would exist such that $l^2 t_1 + g^2 t_2 = h^2$ is true. Comparing degrees we obtain a contradiction. Hence, this example is linked to case (iii) of theorem 31.⋄

Generalized quaternion algebras

The next subsection is dedicated to analyze the structure of the algebras within theorem 31. In addition, we will prove that all three cases lead to pairwise non-isomorphic algebras. Within each case we determine conditions for two algebras of this type being isomorphic.

4.4.1 One component is zero

Let $a \in K \setminus QA(K)$. The next example shows us that the algebra $A(a, 0_K)$ contains two non-conjugated radical complements. This example enhance the topic of counterexamples for the theorem of Wedderburn-Malcev begun in chapter 1 of this work.

Proposition 8 *(counterexample to the conjugacy part of the Wedderburn-Malcev theorem)* Let $A := A(a, 0_K)$. The following statements are valid:

(i) $rad(A) = \langle j, k \rangle_K$

(ii) $A/rad(A)$ is not separable.

(iii) $\langle 1_A, i \rangle_k$ and $\langle 1_A, i + k \rangle_K$ are two different algebra complements of $rad(A)$ in A. Both subalgebras are inseparable field extensions of $K1_A$ which are not conjugated under $1 + rad(A)$.

Proof. By using the multiplication matrix of A we deduce that $\langle j, k \rangle_K$ is a nilpotent ideal and $\langle 1_A, i \rangle_K, \langle 1_A, i+k \rangle_K$ are different subalgebras of A. It is straightforward to prove that these subalgebras are algebra complements of $\langle j, k \rangle_K$ in A. Hence, they are $\mathcal{A}_1$-isomorphic. $\langle 1_A, i \rangle_K = K[i]$ is valid. The polynomial $t^2 + a$ is – based on the assumption of a – irreducible over $K[t]$. Hence, $min_{i,K} = t^2 + a$ is true and $K[i]$ is a field. We deduce $rad(A) = \langle j, k \rangle_K$. $t^2 + a$ possesses the zero i in $K[i]$ of degree 2. Therefor, $t^2 + a$ is inseparable. Based on part (iv) of lemma 1 we conclude that $K[i]$ is a non-separable K-algebra. The proposition is proven because A is commutative.⋄

Theorem 32 *(characterization of isomorphism)* Let $b \in K \setminus QA(K)$. The following statements are equivalent:

(i) $A(a, 0_K) \cong_{\mathcal{A}_1} A(b, 0_K)$

(ii) $a \in \langle 1_K, b \rangle_{QA(K)}$

(iii) $\langle 1_K, a \rangle_{QA(K)} = \langle 1_K, b \rangle_{QA(K)}$

Proof. Let $a, b \in K \setminus QA(K)$ such that

$$dim_{QA(K)}(\langle 1_K, a \rangle_{QA(K)}) = 2 = dim_{QA(K)}(\langle 1_K, b \rangle_{QA(K)})$$

is valid. Straightforward to prove is that parts (ii) and (iii) are equivalent. Let $a \in \langle 1_K, b \rangle_{QA(K)}$. Elements $f, l \in K$ exist such that $a = f^2 + l^2 b$ is true. By using $a \notin QA(K)$ we deduce $l \neq 0_K$. Part (i) is proven by using parts (iii) and (iv) of theorem 29. Let $A(a, 0_K) \cong_{A_1} A(b, 0_K)$ based on an isomorphism γ and $f_1, ..., f_4 \in K$ such that $i\gamma = f_1 1_{A(b,0_K)} + f_2 i + f_3 j + f_4 k$ is true. We calculate

$$a 1_{A(b,0_K)} = i^2 \gamma = (i\gamma)^2 = f_1{}^2 1_{A(b,0_K)} + f_2{}^2 b 1_{A(b,0_K)}.$$

Thus, the theorem is proven. ◇

4.4.2 The third case

Let $a, b, ab \in K \setminus QA(K)$ and no pair exists such that $(g; h) \in K \times K$ mit $g^2 a + h^2 b = 1_K$ is valid.

Proposition 9 *(structural properties)* Let $A := A(a, b)$. *The following statements are valid:*

(i) *A is a splitting field of the polynomial $(t^2 + a 1_A)(t^2 + b 1_A)$ over $K 1_A$ possessing the zeroes i, j in A.*

(ii) *$(K 1_A; A)$ is an inseparable field extension. In particular, A is a non-separable K-algebra.*

(iii) *$(K 1_A; A)$ is no simple field extension. In particular, infinite many intermediate subfields of dimension 2 exist. These are precisely the subalgebras of A different from the zero-space.*

Proof. ad(i): Let $T := \langle 1_A, i \rangle_K$. T is a subalgebra of A. Because of $a \in K \setminus QA(K)$ we deduce that $min_{i,K} = t^2 + a$ is irreducible over K. Hence, $T = K[i]$ is a field. A is commutative, and thus A is a T-algebra. We prove that the polynomial $f := t^2 + b 1_A$ is irreducible over $T[t]$. We assume f would possess a zero s in T. Let $g, h \in K$ such that $s = g 1_A + h i$ is valid. We would deduce $b = g^2 + h^2 a$. Within the case $g \neq 0_K$ this would imply $g^{-2} b + h^2 g^{-2} a = 1_K$ which is a contradiction. Within the case $g = 0_K$ we would calculate $ab = h^2 a^2 = (ha)^2 \in QA(K)$ which is a contradiction, too. Thus, part (i) is proven.

ad(ii): Part (ii) is deductable by using part (i) and part (iv) of lemma 1.

ad(iii): The non-primitivity statement is deductable from the statement $x^2 \in K 1_A$ which implies $dim_K(K[x]) \leq 2$ for all $x \in A$. Let T be a subalgebra of A. A is a field and contains no non-zero nilpotent elements. Therefor, T is semisimple. By using theorem 1 the subalgebra T is unitary. 1_T is an

Generalized quaternion algebras

idempotent von A, and thus $1_T = 0_A$ or $1_T = 1_A$. Within the first case T is the zero-space. By using theorem 1.2.1 in [8] we deduce within the second case that T is an intermediate field of the field extension $(K1_A; A)$.⋄

The next proposition is used for analyzing isomorphism conditions of algebras of the type $A(a,b)$.

Proposition 10 *Let $A := A(a,b)$. The following statements are valid:*

(i) The set $\{1_K, a, b, ab\}$ is $QA(K)$-linear independent.

(ii) The map $a \mapsto a^2$ is a ring monomorphism on A possessing the image $\langle 1_K, a, b, ab \rangle_{QA(K)} 1_A$. In particular, the $\mathbb{Z}$-algebras $\langle 1_K, a, b, ab \rangle_{QA(K)}$ and A are isomorphic.

(iii) $\langle 1_K, a, b, ab \rangle_{QA(K)} \cong_{A_1} A(a^2, b^2, QA(K))$ is valid.

(iv) The Galois group of the field extension $(K1_A; A)$ is trivial.

(v) $QA(K)$ is a subfield of K.

Proof. ad(i)-(iii): Let $f_1, ..., f_4 \in K$ such that $f_1^2 1_K + f_2^2 a + f_3^2 b + f_4^2 ab = 0_K$ is true. We define $a := f_1 1_A + f_2 i + f_3 j + f_4 k$ and calculate $x^2 = (f_1^2 + f_2^2 a + f_3^2 b + f_4^2 ab) 1_A = 0_A$. Based on example 9 the algebra A is a field. Thus, we deduce $x = 0_A$ and $f_i = 0_K$ for all $i \in \underline{4}$. We have proven part (i). Let us define $\gamma : A \longrightarrow A, a \longmapsto a^2$. The function γ is a ring homomorphism. A is a field and $1_A \gamma = 1_A$ is true, and thus γ is a monomorphism. As calculated before $A\gamma = \langle 1_K, a, b, ab \rangle_{QA(K)} 1_A$ is true. Hence, the parts (i) and (ii) are valid. Based on part (i) the set $\{1_K, a, b, ab\}$ is a $QA(K)$-basis of $\langle 1_K, a, b, ab \rangle_{QA(K)}$. If we calculate the multiplication matrix based on this basis and if we compare it to the one of $A(a^2, b^2, QA(K))$, then it is straightforward to prove the isomorphism stated.

ad(iv): Let $\alpha \in Aut_K(A)$. $1_A \alpha = 1_A$ is valid. Let $k_1, ..., k_4 \in K$ such that $i\alpha = k_1 1_A + k_2 i + k_3 j + k_4 k$ is true. We calculate $a 1_A = i^2 \alpha = (i\alpha)^2$, and a comparison of the coefficients lets us deduce $k_1^2 + (k_2+1_K)^2 a + k_3^2 b + k_4^2 ab = 0_K$. We use part (i) and conclude $i\alpha = i$. A similar argumentation shows us $j\alpha = j$.

ad(v): This statement is well-known within the theory of fields.⋄

The following diagram is related to proposition 10. If we apply the statement again and again, then infinite many intermediate fields arise:

$$\begin{array}{c|c}
A(a,b) & \\
K1_{A(a,b)} & \\
<1_{A(a,b)}, a, b, ab>_{Pot(2,K)} & \cong_{\mathcal{A}_1} A(a^2, b^2, Pot(2,K)) \cong_{\mathcal{R}_1} A(a,b) \\
Pot(2,K)1_{A(a,b)} & \\
<1_{A(a,b)}, a^2, b^2, a^2b^2>_{Pot(4,K)} & \cong_{\mathcal{A}_1} A(a^4, b^4, Pot(4,K)) \cong_{\mathcal{R}_1} A(a,b) \\
Pot(4,K)1_{A(a,b)} & \\
<1_{A(a,b)}, a^4, b^4, a^4b^4>_{Pot(4,K)} & \cong_{\mathcal{A}_1} A(a^8, b^8, Pot(8,K)) \cong_{\mathcal{R}_1} A(a,b) \\
Pot(8,K)1_{A(a,b)} & \\
\cdots &
\end{array}$$

Theorem 33 *(characterization of isomorphism)* Let $c, d \in K$, $c, d, cd \in K \setminus QA(K)$ and no pair $(g; h) \in K \times K$ exists such that $g^2 c + h^2 b = 1_K$ is valid. The following statements are equivalent:

(i) $A(a,b) \cong_{\mathcal{A}_1} A(c,d)$

(ii) $a, b \in \langle 1_K, c, d, cd \rangle_{QA(K)}$

(iii) $\langle 1_K, a, b, ab \rangle_{QA(K)} = \langle 1_K, c, d, cd \rangle_{QA(K)}$

Proof. Part (iii) is deductable from part (ii). If part (ii) is valid, then part (iii) can be proven based on remark 10. The implication from part (i) to (ii) is a consequence of (ii) of remark 10: If γ is a $\mathcal{A}_1$-isomorphism between $A(a,b)$ and $A(c,d)$, then

$$a 1_{A(c,d)} = i^2 \gamma = (i\gamma)^2 \in \langle 1_K, c, d, cd \rangle_{QA(K)} \cdot 1_{A(c,d)}$$

is valid. A similar calculation leads to $b \in \langle 1_K, c, d, cd \rangle_{K^2}$. We have to prove the implication from part (ii) to (i). Let $f_1, g_1, ..., f_4, g_4 \in K$ such that $a = f_1^2 + f_2^2 c + f_3^2 d + f_4^2 cd$ and $b = g_1^2 + g_2^2 c + g_3^2 d + g_4^2 cd$ are valid. We define elements in $A(c,d)$ and prove that the set of these elements is a K-basis of $A(c,d)$. We define $i_0 := f_1 1_{A(c,d)} + f_2 i + f_3 j + f_4 k$, $j_0 := g_1 1_{A(c,d)} + g_2 i + g_3 j + f_4 k$ and $k_0 := i_0 j_0$. We calculate

$$\begin{aligned}
i_0^2 &= a 1_{A(c,d)}, \\
j_0^2 &= b 1_{A(c,d)} \text{ and} \\
k_0^2 &= ab 1_{A(c,d)}.
\end{aligned}$$

Let $l_1, l_2, l_3, l_4 \in K$ such that $x := l_1 1_{A(c,d)} + l_2 i_0 + l_3 j_0 + l_4 k_0$ is true. We calculate $x^2 = (l_1^2 + l_2^2 a + l_3^2 b + l_4^2 ab) 1_{A(c,d)}$. For the element $y := l_1 1_{A(a,b)} + l_2 i_0 + l_3 j_0 + l_4 k_0 \in A(a,b)$ we deduce

Generalized quaternion algebras

$$y^2 = (l_1{}^2 + l_2{}^2 a + l_3{}^2 b + l_4{}^2 ab) 1_{A(a,b)} = 0_{A(a,b)}.$$

$A(a, b)$ is – based on example 9 – a field, and thus $y = 0_{A(a,b)}$ is valid. Hence, $l_i = 0_K$ is true for all $i \in \underline{4}$. We focus on the multiplication matrix of $A(c, d)$ based on the K-basis $B := \{1_{A(c,d)}, i_0, j_0, k_0\}$. We calculate (the unit element is not used here):

·	i_0	j_0	k_0
i_0	$a1_{A(c,d)}$	k_0	aj_0
j_0	k_0	$b1_{A(c,d)}$	bi_0
k_0	aj_0	bi_0	$ab1_{A(c,d)}$

Thus, the theorem is proven.◇

Example 10 *(i) Let $K := GF(2)(t_1, t_2)$. $A(t_1, t_2)$ and $A(t_1 + t_2, t_2{}^3)$ are fields which are isomorphic as K-algebras.*

(ii) Let $K := GF(2)(t_1, t_2, t_3)$. $A(t_1, t_2)$ and $A(t_1, t_3)$ are fields which are not isomorphic as K-algebras.

Proof. ad(i): We define $F := GF(2)(t_1)$. $A(t_1, t_2)$ is a field as proven already in part (iii) of remark 24. We calculate $t_1 + t_2, t_2{}^3, t_1 t_2{}^3 + t_2{}^4 \notin QA(K)$. We assume $g, h \in K$ exists such that $g^2(t_1 + t_2) + h^2 t_2{}^3 = 1_K$ is valid. As proven before $g \neq 0_K \neq h$ is valid. Hence, $g, h, f \in F[t_2] \setminus \{0_F\}$ exists such that $g^2(t_1 + t_2) + h^2 t_2{}^3 = f^2$ would be true. $g^2 t_2$ and $h^2 t_2{}^3$ polynomials for which its monomials possess only uneven degrees. $g^2 t_1$ and f^2 are polynomials for which its monomials possess only even degrees. Thus, $g = 0_K = h$ would be valid which is a contradiction. Based on example 9 both K-algebras are fields. Because of $\{t_1 + t_2, t_2{}^3\} \subseteq \langle 1_K, t_1, t_2, t_1 t_2 \rangle_{K^2}$ we deduce part (i) using theorem 33.

ad(ii): We define $F := GF(2)(t_1, t_2)$ and $T := GF(2)(t_1)$. $A(t_1, t_3)$ is a field which was proven already within part (iii) of remark 24. $QA(K) \cap F = QA(F)$ and $QA(F) \cap T = QA(T)$ are valid, and thus $t_1, t_2, t_1 t_2 \notin K^2$ is true. We assume that $A(t_1, t_2)$ is no field. Based on theorem 31 elements $g, h, f \in F[t_3]$ would exist such that $h^2 t_1 + g^2 t_2 = f^2$ is valid. Hence, $0 \neq f, g, h \in F$ would exist such that $g^2 t_1 + h^2 t_2 = f^2$ would be true. But this is a contradiction to part (iii) of remark 24. Based on theorem 9 both K-algebras are fields. A straightforward calculation based on degrees lets us deduce that $t_3 \notin \langle 1_K, t_1, t_2, t_1 t_2 \rangle_{QA(K)}$ is true. Based on theorem 33 part (ii) is proven.◇

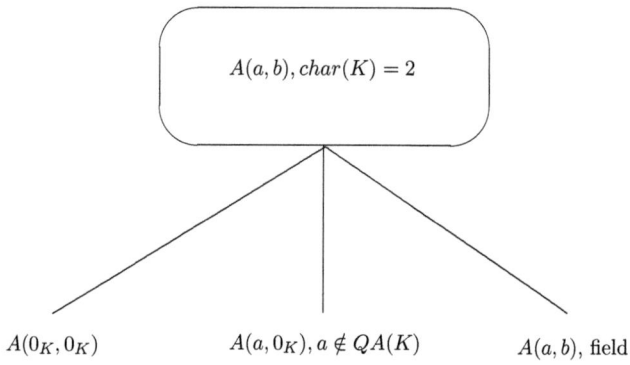

4.5 Exercises

Excercise 262 *Within definition and remark 9 use the cited literature do prove all statements in details.*

Excercise 263 *Within proposition 8 determine and describe all radical complements.*

Excercise 264 *Within remark 23 prove all statements about the class of solvability.*

Excercise 265 *Let a, b two commuting elements of an associative algebra in positive characteristic p. For all $n \in \mathbb{N}$ find a solution for the expression $(a+b)^{p^n}$? Start with the case $p = 2$ and $n = 1$.*

Excercise 266 *Let K be a field with $char(K) = 2$. The set of squares of K is a subfield. Is the same statement true or false for the 2^n-th powers for an arbitrary $n \in \mathbb{N}$?*

Excercise 267 *True or false: The set of squares of $\mathbb{Q}$ is a subfield.*

Excercise 268 *True or false: The set of squares of $\mathbb{R}$ is a subfield,*

Generalized quaternion algebras   149

Excercise 269 *True or false: The set of squares of $\mathbb{C}$ is a subfield.*

Excercise 270 *Let K be a field with uneven characteristic and $a, b \in K$ with $a \neq 0 \neq b$. To which algebra is $A(a, b, K)$ isomorphic?*

Excercise 271 *Let $K := \mathbb{C}$ and $a, b \in K$ with $a \neq 0 \neq b$. To which algebra is $A(a, b, K)$ isomorphic?*

Excercise 272 *Do a research in the literature for all statements within remark 22 and formulate them in details.*

Excercise 273 *True or false: $A(1, -1, \mathbb{R})$ and $A(1, \sqrt{2}, \mathbb{R})$ are isomorphic.*

Excercise 274 *True or false: $A(1, -1, \mathbb{Q})$ and $A(1, 2, \mathbb{Q})$ are isomorphic.*

Excercise 275 *Within remark 10 determine the multiplication matrix in part (iii).*

Excercise 276 *Let K be a field with 2 elements and $a, b \in K$. How many non-isomorphic algebras $A(a, b, K)$ do exist? For each algebra determine its nilradical and the unique radical complement. Why is the radical complement unique? Draw a Hasse diagram for each algebra including the results of this exercise.*

Excercise 277 *Solve exercise 276 based on field with 4 elements.*

Excercise 278 *Generalize the statement in exercise 277 to an arbitrary field possessing 2^n elements.*

Excercise 279 *Let $K := GF(2)(t)$ and $a, b \in \{0, 1, t, t^2, t^3, t+t^2, t+t^3, t^2+t^3\}$. Analyze how many pairwise non-isomorphic algebras $A(a, b, K)$ arise.*

Excercise 280 *Let $K := GF(2)(t_1, t_2)$ and $a, b \in \{0, 1, t_1, t_2, t_1t_2, t_1^2, t_2^2, t_1+t_2, t_1^3, t_2^3\}$. Analyze how many pairwise non-isomorphic algebras $A(a, b, K)$ arise.*

Excercise 281 *Let K be a field, $\mathrm{char}(K) = 2$ and D a 4-dimensional associative central K-division algebra. It is well-known that elements $a, b \in K \setminus \{0\}$ and a K-basis $\{1, i, j, k\}$ exist possessing the following structure constants:*

·	1	i	j	k
1	1	i	j	k
i	i	$a1$	k	aj
j	j	$k+i$	$j+b1$	$b1$
k	k	$a(j+1)$	$k+bi$	$ab1$.

These algebras are the so-called quaternion algebras in characteristic two. The aim of this exercise is to do the same analysis for these kind of algebras as done in chapter 4 by assuming $a = 0$ or $b = 0$. In addition, analyze the change of the characteristic to not two!

Chapter 5
Commutative algebras

One main class of examples of commutative algebras are centers of (non-commutative) arbitrary algebras. The first section is focussed on the question to determine the unique radical complement of the center by using the algebra containing it (idea of compatibility). Afterwards the inner structure of commutative algebras are analyzed by using an intrinsic description of the unique radical complement. Related sets of diagonalizable and splitting elements are linked to this unique radical complement. The results are applied to the algebras arising within chapter 4, to commutative group algebras and to arbitrary solvable algebras to describe their radical complements. Within this chapter we answer the decomposition question for an element based on the generalized Jordan decomposition for commutative and for solvable algebras. As a consequence, we describe a radical complement (and the radical) by using the decomposition of single elements for commutative and solvable algebras: we call this construction the bottom-up construction (instead of the presented top-down calculation within chapter 2). All results are transferred to non-unital algebras using the adjunction of an unit.

5.1 Compatibility with the center

We start the analysis of determining the unique radical complement of the center.

Remark 25 *(center of a separable algebra) Let K be a field and A an associative separable K-algebra. Every K-subalgebra of $Z(A)$ is separable. In particular, $Z(A)$ is separable.*

Proof. We start the proof by deducing that $Z(A)$ is separable. By using statement (iv) of theorem 1 an element $r \in \mathbb{N}$ and associative finite-dimensional unitary simple K-algebras A_i ($i \in \underline{r}$) exist such that $A \cong_{A_1}$

$\bigoplus_{i=1}^{r} A_i$ is valid and for every $i \in \underline{r}$ the field extension $(K1_{A_i}; Z(A_i))$ is separable. Part (iii) of corollary 1 lets us deduce that for every $i \in \underline{r}$ the set $Z(A_i)$ is a separable $K1_{A_i}$-algebra, and hence is a separable K-algebra, too. By using part (ii) of proposition 2 the set $Z(A)$ is a separable K-algebra. Let T be a K-subalgebra of $Z(A)$. Part (i) of corollary 1 is used to prove that T is finite-dimensional. By using part (i) of theorem 26 we deduce that $rad(T) = \{0_A\}$ is valid. Hence, T is semisimple. $Z(A)$ is commutative, and thus solvable, too, and by using part (ii) of theorem 26 the remark is proven.⋄

Now we are almost ready to apply the theorem of Wedderburn-Malcev to central subalgebras. Within the next proposition we are checking its assumptions.

Proposition 11 *Let K be a field, A a finite-dimensional associative K-algebra, $A/rad(A)$ separable and T a K-subalgebra of $Z(A)$. The following statements are valid:*

(i) $rad(T) = rad(A) \cap T = rad(Z(A)) \cap T$
 In particular, $rad(Z(A)) = rad(A) \cap Z(A)$ *is valid.*

(ii) $T/rad(T)$ *is separable.*
 In particular, $Z(A)/rad(Z(A))$ *is separable.*

Proof. We start the proof by deducing the statements for $Z(A)$. $rad(A) \cap Z(A)$ is a nilpotent ideal of $Z(A)$. In addition, $Z(A)/(rad(A) \cap Z(A))$ is $\mathcal{A}_1$-isomorphic to a K-subalgebra of the center of $A/rad(A)$. By using remark 25 we deduce the separability of $Z(A)/(rad(A) \cap Z(A))$. In particular, part (i) and (ii) are valid for $Z(A)$.
Let T be a K-subalgebra of $Z(A)$. $Z(A)$ is commutative, and thus it is also solvable. By using theorem 26 we have proven the proposition.⋄

We know that within the context of proposition 11 a central subalgebra possesses exactly one radical complement because it is commutative and all radical complement are conjugated (see part (vi) of theorem 16 and theorem 15). The following remark is needed for its description.

Remark 26 *Let A be an associative K-algebra, T a semisimple K-subalgebra of A and $X := \bigcap_{r \in rad(A)} T^{(r)}$ the intersection of all conjugates of T under the action of $rad(A)$. The statement $X \subseteq C_A(rad(A))$ is valid.*

Proof. X is invariant under the conjugation of $rad(A)$. Let $x \in X \subseteq T$, $r \in rad(A)$. $x^{(r)} \in X \subseteq T$ is valid. By using part (ii) of theorem 3 an element $s \in rad(A)$ exists such that $x^{(r)} = x + s$ is valid. T is semisimple, and thus $s \in rad(A) \cap T \subseteq rad(T) = \{0_T\}$ is true, and we deduce $x^{(r)} = x$. By using part (ii) of theorem 3 the remark is proven.⋄

Commutative algebras

Theorem 34 *(radical complement of central subalgebras) Let K be a field, A a finite-dimensional associative K-algebra, $A/rad(A)$ separable, D a radical complement of A and T a central K-subalgebra of A. The following statements are valid:*

(i) $D \cap T$ is the unique radical complement of T.

(ii) $D \cap T$ is the unique maximal semisimple and semisimple K-subalgebra of T.

Proof. Let S be the unique radical complement of T (see part (ii) of proposition 11, theorem 13, part (vi) of calculation 16 and theorem 15). We start the proof by deducing part (ii) for S. By using part (ii) of proposition 11 the K-subalgebra S is separable, and thus it is – based on part (i) of corollary 1 – also semisimple. Let H be a semisimple K-subalgebra of T. Part (ii) of proposition 11 and part (ii) of theorem 26 let us deduce the separability of H. T is central, and thus by using part (i) of corollary 4 the statement $H \subseteq S$ is true. We have proven part (ii) for S.

We have to finish the proof by deducing $S = D \cap T$. $D \cap T$ is a central K-subalgebra, and thus part (i) of proposition 11 lets us deduce that $rad(D \cap T) = rad(A) \cap (D \cap T)$ is valid. This statement implies that $D \cap T$ is semisimple and that $D \cap T \subseteq S$ is valid. By using part (ii) of proposition 11 the set S is a separable K-subalgebra of A. S is central, and by using part (i) of corollary 4 we deduce $S \subseteq D$. ◇

Corollary 6 *(intersection of all radical complements within solvable algebras) Let K be a field, A a finite-dimensional associative solvable K-algebra, $A/rad(A)$ separable, D a radical complement of A and X the intersection of all radical complements of A. The following statements are valid:*

(i) X is the unique radical complement of $Z(A)$.

(ii) X is the unique maximal semisimple and semisimple K-subalgebra of $Z(A)$.

(iii) If $Z(A)$ is semisimple, then $X = Z(A)$ is valid.

(iv) If A is unitary and central[1], then $X = K1_A$ is valid.

Proof. ad(i) and (ii): By using remark 25 the statement $X \subseteq C_A(rad(A))$ is valid. A is solvable, and hence D is commutative. Therefor, we deduce

$$X \subseteq C_A(rad(A)) \cap C_A(D) \subseteq C_A(rad(A) + D) = Z(A).$$

---
[1] Central algebras are associative unitary algebras for which the center is exactly the K-linear span of the unit element.

Thus, $X = X \cap Z(A)$ is valid. By using theorem 34 we have proven parts (i) and (ii).

ad(iii): This statement is a consequence of part (ii).

ad(iv): Because of $K1_A \subseteq X$ (see remark 4) this part is deductable from statement (ii).◇

Examples 4 (i) Let $n \in \mathbb{N}$ and D_n the Solomon algebra. For a deep insight of the structure of the Solomon algebra the reader may study the dissertation of T. Bauer (see [3]). By using lemma 3.4 in [3] we know that D_n is solvable. Theorem 3.6 and lemma 3.4 in [3] let us deduce that $Z(D_n)$ is contained in a radical complement of D_n. Thus, the center is exactly the intersection of all radical complements of D_n.

(ii) Within example 3.4 we have proven that the center is exactly set K-linear span of the unit element. Thus, the center is exactly the intersection of all radical complements (see part (iii) of corollary 6).

(iii) The same result as stated in part (ii) is true in the more general context of algebras of lower and upper triangular matrices over a field K because they are central.◇

5.2 The subalgebra of fully separable elements

Now we focus on the inner structure of commutative algebras. We start the analysis by defining several properties of elements which are all linked to the unique radical complement or the nilradical of commutative algebras.

Definitions 5 *(semisimple, nilpotent, separable, fully separable, diagonalizable, splitting)* Let K be a field and $f \in K[t]$. f is called semisimple (or also squarefree) in $K[t]$, if f is the product of pairwise distinct irreducible polynomials of $K[t]$. If all of these factors are linear, then f is called diagonalizable in $K[t]$. f is called fully separable in $K[t]$, if f is semisimple and separable in $K[t]$. If $\prod_{i=1}^{n} f_i^{s_i}$ is the decomposition of f into irreducible polynomials in $K[t]$, then we define $halb(f) := \prod_{i=1}^{n} f_i$ and $max(f) := max\{s_1, \cdots, s_n\}$. f is called splitting in $K[t]$, if $halb(f)$ is diagonalizable over $K[t]$. f is called nilpotent, if $f = t^n$ is valid for an element $n \in \mathbb{N}$. The definition of a separable polynomial is used as defined usually within the context of field theory.◇

Definitions 6 Let K be a field, A an associative unitary K-algebra and $a \in A$ an algebraical element over K. a is called semisimple, nilpotent, separable,

Commutative algebras 155

fully separable, diagonalizable resp. splitting over K if $min_{a,K}$ is semisimple, nilpotent, separable, fully separable, diagonalizable resp. splitting over $K[t]$. By $H(A)$, $Nil(A)$, $Sep(A)$, $VSep(A)$, $D(A)$ resp. $ZF(A)$ we denote the set of all semisimple, nilpotent, separable, fully separable, diagonalizable resp. splitting elements of A over K.
If K is a field and $f \in K[t], f \neq 0_K$, then $grad(f)$ is the degree of the polynomial f and (f) the K-ideal $fK[t]$ of $K[t]$.⋄

Remark 27 (i) Let K be a field and A be an associative unitary K-algebra. $Sep(A)$ is containing every nilpotent element of A. The intersection of $VSep(A)$ with the set of all nilpotent elements of A is exactly $\{0_A\}$. In addition, $K1_A \subseteq D(A) \subseteq VSep(A) \subseteq H(A) \cap Sep(A)$ and $rad(A) \cup D(A) \subseteq ZF(A) \subseteq Sep(A)$ are valid.

(ii) Let $(K; L)$ be a field extension. Every element of the K-algebra L is separable over K if and only if it is fully separable over K. Thus, $Sep(L) = VSep(L)$ is valid.⋄

Example 11 Let K be a field.

(i) We focus on the K-algebra $A := A(1_K, 0_K, K)$ in the case $char(K) \neq 2$ (see the section 23 of chapter 4). The statement $rad(A) = \langle j, k \rangle_K$ is valid. Because of $i^2 = 1_A$ we deduce $min_{i,K} = t^2 - 1_K = (t+1_K)(t-1_K)$. Hence, $i \in VSep(A)$ is valid. In addition, $(1_A + j)^{-1} = 1_A - j$ is true. Thus, $i^{1_A+j} = i + 2_K k$ is fully separable over K. If $VSep(A)$ would be a K-space, then k would be fully separable over K. By definition of A the element k is nilpotent. Based on part (i) of remark 27 this would imply $k = 0_A$ which is a contradiction. Hence, $VSep(A)$ is no K-space.

(ii) Let $char(K) = 2$ and $A := A(0_K, 0_K, K)$ (see section 21 of chapter 4). $rad(A) = \langle i, j, k \rangle_K$ is true. Let $x \in A$ and $f_1, ..., f_4 \in K$ such that $x = f_1 1_A + f_2 i + f_3 j + f_4 k$ is valid. We calculate $x^2 = f_1^2 1_A$ and $(x + f_1 1_A)^2 = 0_A$. Hence, we have determined $VSep(A) = K1_A$ and $A = Sep(A) = VSep(A) + rad(A)$.

(iii) Let $char(K) = 2$, $a \in K \setminus QA(K)$ and $A := A(a, 0_K, K)$ (see examples 8 and 24). $rad(A) = \langle j, k \rangle_K$ is valid. Let $x \in A$ and $f_1, ..., f_4 \in K$ such that $x = f_1 1_A + f_2 i + f_3 j + f_4 k$ is true. We calculate $x^2 = f_1^2 1_A + f_2^2 a$. If $f_2 = 0_K$ is valid, then x is separable over K. In this case x is fully separable over K if and only if $x \in K1_A$ is true. If $f_2 \neq 0_K$ and $x \notin K1_A$ are valid, then $min_{x,K} = t^2 + f_1^2 + f_2^2 a \notin QA(K)[t]$ is true. In this case x is not separable over K. Thus, $VSep(A) = K1_A$ and $Sep(A) = VSep(A) + rad(A)$ are valid.

(iv) Let $char(K) = 2$ and $a, b \in K$ such that $A := A(a, b, K)$ is a field (see examples 9 and 24 in chapter 4). If $x \in A$, then $k \in K$ exists such that $x^2 = k1_A$ is true. If $x \notin K1_A$, then $min_{x,K} = t^2 - k$ is true. Within the field A we calculate $t^2 - k1_A = (t + a)^2$. Thus, $Sep(A) = VSep(A) = K1_A$ is valid.⋄

Lemma 7 *Let K be a field and A an associative unitary K-algebra. The following statements are valid:*

(i) Let $f \in K[t]$. f is fully separable in $K[t]$ if and only if $K[t]/(f)$ is separable as K-algebra.

(ii) Let $a \in A$. $a \in VSep(A)$ is valid if and only if $K[a]$ is separable as K-algebra. In particular, a is fully separable if and only if $min_{a,K}$ is fully separable.

(iii) If A is commutative and separable as K-algebra, then $A = VSep(A)$ is valid.

Proof. ad(i): If f is fully separable in $K[t]$, then f is by definition semisimple in $K[t]$. If the K-algebra $K[t]/(f)$ is separable, then part (i) of corollary 1 let us deduce that it is semisimple. It is well-known that the algebra is semisimple if and only f is semisimple in $K[t]$ by the Chinese Remainder theorem. Thus, we can assume that f is semisimple in $K[t]$. Let $n \in \mathbb{N}$ and $f_1, ..., f_n$ pairwise distinct irreducible polynomials in $K[t]$ such that $f = \prod_{i=1}^{n} f_i$ is valid. By using the Chinese Remainder theorem we deduce that $K[t]/(f) \cong_{A_1} \bigoplus_{i=1}^{n} K[t]/(f_i)$ is true. Because of part (ii) of proposition 2 the algebra $\bigoplus_{i=1}^{n} K[t]/(f_i)$ is separable if and only if for all $i \in \underline{n}$ the algebra $K[t]/(f_i)$ is separable. Part (iii) of corollary 1 let us deduce part (i).

ad(ii): If a is algebraical over K, then $K[a] \cong_{A_1} K[t]/(min_{a,K})$ is true. By using this and part (i) part (ii) is proven because – based on part (i) of corollary 1 – every separable K-algebra is finite-dimensional and thus algebraical, too.

ad(iii): Let $a \in A$. By using part (ii) of remark 18 the algebra $K[a]$ is a separable subalgebra of A. We conclude that part (iii) is valid using part (ii).⋄

Within the following theorem we apply another property of separable algebras: the tensor product of two separable algebras is separable. This statement is used within several proofs of this work. (For a proof see e.g.

Commutative algebras   157

the corollary within section 10.5 of the text book [35]. The reader may prove
this fact also within the exercises (see exercise 286).).

Theorem 35 *(the subalgebra of fully separable elements)* Let K be a field
and A a commutative associative unitary K-algebra. $VSep(A)$ is a K-subalgebra of A.

Proof. It is straightforward to prove $\{1_K, 0_K\} \subseteq VSep(A)$. Let $a, b \in VSep(A)$. Based on part (iii) of lemma 7 we have to prove only that the commutative K-algebra $K[a, b]$ is separable. $K[a]$ and $K[b]$ are K-subalgebras of $K[a, b]$ such that $\langle K[a]K[b]\rangle_K = K[a, b]$ is valid and their elements are commutating pairwise. Hence, $K[a, b]$ is a epimorphic image of the tensor product $K[a]\otimes_K K[b]$. By using part (ii) of lemma 7 the K-subalgebras $K[a]$ and $K[b]$ are separable. The corollary in chapter 10.5 of [35] let us deduce that $K[a]\otimes_K K[b]$ is separable, too. Part (i) of proposition 2 implies that $K[a, b]$ is separable, too. ◇

5.3 The context of the Wedderburn-Malcev theorem

Theorem 36 *(separability and fully separability in commutative algebras)*
Let K be a field and A a finite-dimensional associative commutative unitary
K-algebra. The following statements are valid:

(i) $Sep(A) = VSep(A) \oplus_K rad(A)$ is valid. In particular, $Sep(A)$ is a
    K-subalgebra of A.

(ii) $VSep(A)$ is a separable K-algebra.

(iii) $VSep(A)$ is the unique algebra complement of $rad(Sep(A)) = rad(A)$
    in $Sep(A)$. In particular, $Sep(A)/rad(A)$ is separable.

(iv) A is separable if and only if $A = VSep(A)$ is true.

(v) The following statements are equivalent:

   (a) $A/rad(A)$ is separable.
   (b) $A = Sep(A)$ is valid.
   (c) $VSep(A)$ is a radical complement in A.
   (d) $VSep(A)$ is the unique radical complement in A.

Proof. ad(i): By using remark 27 we deduce that $rad(A) \cup VSep(A) \subseteq Sep(A)$ and $rad(A) \cap VSep(A) = \{0_A\}$ are true. Theorem 35 lets us conclude that $VSep(A)$ is a K-subalgebra of A. Hence, the add-on is valid because the sum of an ideal and a subalgebra is a subalgebra. We have only to prove that $rad(A) + VSep(A) \subseteq Sep(A)$ and $Sep(A) \subseteq rad(A) + VSep(A)$ are valid. We start the argumentation by proving the second inclusion. Let $a \in Sep(A)$. $min_{a,K}$ is separable in $K[t]$, and based on part (i) of lemma 7 the algebra $K[a]/rad(K[a])$ is separable. Theorem 13 is used to prove that $rad(K[a])$ possesses an algebra complement X in $K[a]$. Because of part (iii) of lemma 7 we deduce $X \subseteq VSep(A)$. $rad(K[a]) = rad(A) \cap K[a]$ is true, and thus the inclusion is proven. Now we prove the first inclusion. Let $r \in rad(A)$ and $v \in VSep(A)$. $B := K[r+v]$ is a K-subalgebra of $T := rad(A) \oplus_K K[v]$. v is fully separable over K, and thus – based on part (ii) of lemma 7 – the set $K[v]$ is a separable K-algebra which is – because of part (vi) of section 16 – the unique algebra complement of $rad(A)$ in T. The theorems 26 and 27 let us deduce that $K[v] \cap B$ is algebra complement of $rad(B) = rad(T) \cap B$ in B. Because of part (ii) of remark 18 the algebra $K[v] \cap B$ is separable and the statement $K[v] \cap B \cong_{A_1} K[t]/(halb(min_{r+v,K}))$ is true. Part (i) is now a consequence of lemma 7.

ad(ii): Let $n \in \mathbb{N}$ and $B := \{b_1, ..., b_n\}$ a K-basis of $VSep(A)$. Based on theorem 35 the statement $VSep(A) = \langle B \rangle_K = K[B]$ is valid. The algebras $K[b_i]$ ($i \in \underline{n}$) are pairwise commutating subalgebras of $VSep(A)$ such that $VSep(A) = \langle \prod_{i=1}^{n} K[b_i] \rangle_K$ is valid. Hence, $VSep(A)$ is a homomorphic image of the tensor product $\bigotimes_{i=1}^{n} K[b_i]$. By using part (ii) of lemma 7 for every $i \in \underline{n}$ the set $K[b_i]$ is a separable K-algebra. The corollary in section 10.5 of the textbook [35] lets us deduce that $\bigotimes_{i=1}^{n} K[b_i]$ is separable, too. Because of part (i) of proposition 2 also $VSep(A)$ is separable.

ad(iii): Based on part (i) the statement $rad(Sep(A)) = rad(A)$ is valid. By using parts (i) and (ii) we deduce the separability of $Sep(A)/rad(A)$. Part (iii) is now a consequence of part (i) and part (vi) of corollary 16.

ad(iv): If $A = VSep(A)$ is true, then A is separable based on part (ii). Let A be separable. Part (iii) of lemma 7 is used to prove that every element of A is fully separable over K.

ad(v): (a) $\Rightarrow$ (d): Let $A/rad(A)$ be separable. Based on corollary 16 the set $rad(A)$ possesses exactly one algebra complement T in A. This complement is separable. Hence, based on part (iii) of lemma 7 the statement $T \subseteq VSep(T) \subseteq VSep(A)$ is valid. Based on part (i) we deduce $VSep(A) \cap rad(A) = \{0_A\}$. Dedekind's law is used to deduce $VSep(A) = T$.

(d) $\Rightarrow$ (c): This implication is straightforward to prove.

Commutative algebras 159

$(c) \Rightarrow (b)$: This statement is a consequence of (i).

$(b) \Rightarrow (a)$: This implication is deductable by part (ii). ⋄

Corollary 8 *(characterization of separable elements) Let K be a field, A an associative unitary K-algebra and $a \in A$ algebraical over K. The following statements are equivalent:*

(i) a is separable over K.

(ii) $K[a]/rad(K[a])$ is separable.

(iii) $Sep(K[a]) = K[a] = rad(K[a]) \oplus_K VSep(K[a])$ is valid.

Proof. The statements (ii) and (iii) are equivalent based on parts (i) and (iv) of theorem 36. The implication from (iii) to (i) is straightforward to prove. Let part (i) be valid. $Sep(K[a])$ is – based on part (i) of theorem 36 – a K-subalgebra of $K[a]$, and thus $Sep(K[a]) = K[a]$ is true. The rest of the proof is deductable by part (i) of theorem 36. ⋄

Corollary 9 *(properties of the subalgebras $Sep(A)$ and $VSep(A)$) Let K be a field and A an associative commutative finite-dimensional unitary K-algebra. The following statements are valid:*

(i) $VSep(A)$ is the unique maximal semisimple and separable K-subalgebra of $Sep(A)$.

(ii) $VSep(A)$ is the unique maximal separable algebra of A.

(iii) $Sep(A)$ is the unique maximal unitary K-subalgebra of A possessing a separable factor algebra by its nilradical.

Proof. ad(i) and (ii): By using theorem 36 the set $VSep(A)$ is a separable K-algebra. In particular, $VSep(A)$ is – based on part (i) of corollary 1 – semisimple. Part (ii) of theorem 26 and parts (i) and (ii) of theorem 36 let us deduce that every semisimple K-subalgebra of $Sep(A)$ is separable. Let T be a separable K-subalgebra of A. Based on part (iv) of theorem 36 we conclude $T = VSep(T) \subseteq VSep(A)$.

ad(iii): Parts (i) and (ii) of theorem 36 let us deduce that $Sep(A)$ possesses the desired properties. Let T be an unitary K-subalgebra of A possessing a separable factor algebra by its nilradical. We use part (iv) of theorem 36 to deduce that $T = Sep(T) \subseteq Sep(A)$ is valid. ⋄

Within the following example we calculate a radical complement for an algebra which arise in a linear algebraic context.

Example 12 Let $K := \mathbb{R}$, $a \in End_K(K^4)$ such that $min_{a,K} = (t^2+1)^2$ and $A := K[a]$ are valid. K is a perfect field and A is an associative unitary K-algebra possessing the dimension $dim_K(A) = grad(min_{a,K}) = 4$. By using the Chinese-Remainder theorem we deduce $rad(A) = (t^2+1)/((t^2+1)^2)$ and $A/rad(A) \cong_{A_1} \mathbb{C}$. In particular, based on theorem 36 the algebra $VSep(A)$ is the unique radical complement in A. $VSep(A)$ is two-dimensional, and thus we focus on two linear independent and fully separable elements over K in A. Straightforward to prove is that $1_A \in VSep(A)$ and $min_{a^2,K} = t^2+2t+2$ are valid. The latter polynomial is irreducible over the perfect field K. Thus, $VSep(A) = \langle 1_A, a^2 \rangle_K$ is valid. ⋄

An interesting question is for which algebras A the set $VSep(A)$ is a radical complement. This topic will be analyzed within the last section of this work. In addition, within that section the set $VSep(A)$ will be used to describe the radical complements of solvable associative algebras A.

Corollary 10 *(solvable algebras and separable radical factor structure)* Let K be a field and A a finite-dimensional associative solvable unitary K-algebra. The following statements are valid:

(i) $A/rad(A)$ is separable.

(ii) $A = Sep(A)$

(iii) $A/rad(A) = VSep(A/rad(A))$.

Proof. The conditions (i) and (iii) are equivalent because $A/rad(A)$ is commutative (see theorem 5.3). Let $A/rad(A)$ be separable and $a \in A$. We focus on the subalgebra $K[a]$. Its radical factor algebra is separable based on theorem 26. Again using theorem 5.3 the element a is separable. Now we assume that every element of A is separable. Let $a \in A$. We have to prove that $a + rad(A)$ is fully-separable (condition (iii)). By using corollary 8 we can decompose a into $r+t$ such that $r \in rad(K[a])$ and t is fully-separable. A is solvable, and thus $r \in rad(A)$ is valid. We derive $a+rad(A) = t+rad(A)$. Again by using corollary 8 we have to prove that $K[t+rad(A)]$ is separable. One possible argument is related to exercise 283. If t is fully-separable, then $t+rad(A)$ is fully-separable, too, because both elements possess the same minimum polynomial. An alternative approach is that $K[t+rad(A)] = (K[t]+rad(A))/rad(A)$. The latter algebra is isomorphic to $K[t]$ if t is fully separable because in that case $K[t] \cap rad(A) = 0$ is true. ⋄

5.4 A generalized Jordan decomposition

5.4.1 The construction of the decomposition

For commutative associative algebras we have characterized the radical complement by using the set of fully separable elements. Within this section we want to represent a separable element as a sum of a nilpotent and fully separable element which are commuting with each other. This representation can be calculated constructively based on the minimal polynomial of the separable element. The analysis begins with the study of finite-dimensional associative unitary local algebras. Such algebras possess a factor algebra modulo the nilradical which is isomorphic to a division algebra. In the case of a commutative algebra this division algebra is a field.

Proposition 12 *(the local situation) Let K be a field, $n \in \mathbb{N}$, $f \in K[t]$ irreducible and separable in $K[t]$ and $A := K[t]/(f^n)$. $rad(A) = (f)/(f^n)$ is valid and $A/rad(A)$ is A_1-isomorphic to the field $K[t]/(f)$. A is a local K-algebra. Based on part (i) of lemma 7 we deduce that $A/rad(A)$ is separable. Theorem 36 is used to prove that $A = Sep(A) = rad(A) \oplus_K VSep(A)$ is valid and that $VSep(A)$ is the unique radical complement in A. We present three descriptions for $VSep(A)$, and the third one is the most important one concerning our analysis. The following statements are valid:*

(i) *An element $g \in K[t]$ exists such that $f^n \mid f(g)$ and $VSep(A) = K[g + (f^n)]$ are valid.*

(ii) $VSep(A) = E(A) \dot{\cup} \{0_A\}$

(iii) $VSep(A) = \{R \mid \exists s \in K[t] : R = s + (f^n) \wedge (s = 0_K \vee grad(s) < grad(f))\}$

(iv) *Let $g \in K[t]$. If $(r; s) \in K[t] \times K[t]$ is the pair such that $g = rf + s$ and $s = 0_K$ or $grad(s) < grad(f)$ are valid, then $g + (f^n) = (rf + (f^n)) + (s + (f^n))$, $rf + (f^n) \in rad(A)$ and $s + (f^n) \in VSep(A)$ are true.*

Proof. ad(i): Based on Kronecker's construction[2] the polynomial f possesses a zero $g + (f^n)$ in $VSep(A)$. $g + (f^n)$ is a zero of f in $VSep(A)$, and

---

[2] Leopold Kronecker (born 7th December 1823, died 29 December 1891) was a German mathematician who worked on number theory, algebra and logic. He criticized Georg Cantors work on set theory, and was quoted by Weber (1893) as having said, 'Die ganzen Zahlen hat der liebe Gott gemacht, alles andere ist Menschenwerk' ('God made the integers, all else is the work of man.'). Kronecker was a student and lifelong friend of Ernst Kummer. Leopold Kronecker was born on 7th December 1823 in Liegnitz, Prussia (now Legnica, Poland) in a wealthy Jewish family. His parents, Isidor and Johanna, took care of their childrens education and provided them with private tutoring at home. Leopold's younger brother Hugo Kronecker would also follow a scientific path, later becoming a notable physiologist. Kronecker then went to the Liegnitz Gymnasium where he was interested in a wide range of topics including science, history and philosophy, while also practicing gymnastics and swimming. At the gymnasium he was taught by Ernst Kummer, who noticed and encouraged the boy's interest in mathematics. In 1841 Kronecker became a student at the University of Berlin where his interest did not immediately focus on mathematics, but rather spread over several subjects including astronomy and philosophy. He spent the summer of 1843 at the University of Bonn studying astronomy and between 1843 and 1844 at the University of Breslau following his former teacher Kummer. Back in Berlin, Kronecker studied mathematics with Peter Gustav Lejeune Dirichlet and

thus $f^n \mid f(g)$ is true. We use the irreducibility of f in $K[t]$ to obtain part (i).

ad(ii): This statement is straightforward to prove by using the fact that

---

in 1845 defended his dissertation in algebraic number theory written under Dirichlet's supervision. After obtaining his degree, Kronecker did not follow his interest in research on an academic career path. He went back to his hometown to manage a large farming estate built up by his mother's uncle, a former banker. In 1848 he married his cousin Fanny Prausnitzer, and the couple had six children. For several years Kronecker focused on business, and although he continued to study mathematics as a hobby and corresponded with Kummer, he published no mathematical results. In 1853 he wrote a memoir on the algebraic solvability of equations extending the work of Evariste Galois on the theory of equations. Due to his business activity, Kronecker was financially comfortable, and thus he could return to Berlin in 1855 to pursue mathematics as a private scholar. Dirichlet, whose wife Rebecka came from the wealthy Mendelssohn family, had introduced Kronecker to the Berlin elite. He became a close friend of Karl Weierstrass, who had recently joined the university, and his former teacher Kummer who had just taken over Dirichlet's mathematics chair. Over the following years Kronecker published numerous papers resulting from his previous years' independent research. As a result of this published research, he was elected a member of the Berlin Academy in 1861. Although he held no official university position, Kronecker had the right as a member of the Academy to hold classes at the University of Berlin and he decided to do so, starting in 1862. In 1866, when Riemann died, Kronecker was offered the mathematics chair at the University of Göttingen (previously held by Carl Gauss and Dirichlet), but he refused, preferring to keep his position at the Academy. Only in 1883, when Kummer retired from the University, was Kronecker invited to succeed him and became an ordinary professor. Kronecker was the supervisor of Kurt Hensel, Adolf Kneser, Mathias Lerch, and Franz Mertens, amongst others. His philosophical view of mathematics put him in conflict with several mathematicians over the years, notably straining his relationship with Weierstrass, who almost decided to leave the University in 1888. Kronecker died on 29 December 1891 in Berlin, several months after the death of his wife. In the last year of his life, he converted to Christianity. He is buried in the Alter St Matthäus Kirchhof cemetery in Berlin-Schöneberg, close to Gustav Kirchhoff. An important part of Kronecker's research focused on number theory and algebra. In an 1853 paper on the theory of equations and Galois theory he formulated the Kronecker-Weber theorem, without however offering a definitive proof (the theorem was proved completely much later by David Hilbert). He also introduced the structure theorem for finitely-generated Abelian groups. Kronecker studied elliptic functions and conjectured his 'liebster Jugendtraum' ('dearest dream of youth'), a generalization that was later put forward by Hilbert in a modified form as his twelfth problem. In an 1850 paper, On the Solution of the General Equation of the Fifth Degree, Kronecker solved the quintic equation by applying group theory (though his solution was not in terms of radicals: that was already proven impossible by the Abel-Ruffini theorem). In algebraic number theory Kronecker introduced the theory of divisors as an alternative to Dedekind's theory of ideals, which he did not find acceptable for philosophical reasons. Although the general adoption of Dedekind's approach led Kronecker's theory to be ignored for a long time, his divisors were found useful and were revived by several mathematicians in the 20th century. Kronecker also contributed to the concept of continuity, reconstructing the form of irrational numbers in real numbers. In analysis, Kronecker rejected the formulation of a continuous, nowhere differentiable function by his colleague, Karl Weierstrass. Also named for Kronecker are the Kronecker limit formula, Kronecker's congruence, Kronecker delta, Kronecker comb, Kronecker symbol, Kronecker product, Kronecker's method for factorizing polynomials, Kronecker substitution, Kronecker's theorem in number theory, Kronecker's lemma, and Eisenstein-Kronecker numbers.

Commutative algebras   163

A is local.

ad(iii): Let $g \in K[t]$ such that $g = 0_K$ or $grad(g) < grad(f)$ is true. If $g + (f^n) \notin VSep(A)$ is valid, then we use the locality of A to prove $g + (f^n) \in rad(A) = (f)/(f^n)$. Hence, $f \mid g$ is valid which is a contradiction to the degrees of f and g.
Let $s + (f^n) \in VSep(A)$ and $u, r \in K[t]$ such that $s = uf + r$ and $r = 0_K$ or $grad(r) < grad(f)$ are valid. As proven before we derive $r+(f^n) \in VSep(A)$. Hence, $uf + (f^n) = 0_A$ and $s + (f^n) = r + (f^n)$ are true.

ad(iv): This part is a consequence of statement (iii). ◇

Now we are almost ready to construct the generalized Jordan decomposition. For this we need the next definition.

Definition 8 *(nilpotent and fully separable part)* (i) Let K be a field and A an associative K-algebra. Based on corollary 8 for every $a \in Sep(A)$ exactly one pair $(a_{nil}; a_{vsep}) \in rad(K[a]) \times VSep(K[a])$ exists such that $a = a_{nil} + a_{vsep}$ is valid. We define

$$Z_A : Sep(A) \longrightarrow A \times A, a \longmapsto (a_{nil}; a_{vsep}).$$

(ii) Let $n \in \mathbb{N}$ and $A_1, ..., A_n$ associative K-algebras. We define

$$S(A_i, n) : \bigoplus_{i=1}^{n}(A_i \times A_i) \longrightarrow \bigoplus_{i=1}^{n} A_i \times \bigoplus_{i=1}^{n} A_i$$
$$((a_1; b_1), ..., (a_n; b_n)) \longmapsto ((a_1, ..., a_n); (b_1, ..., b_n)). \diamond$$

Theorem 37 *(construction of the nilpotent and fully separable part)* Let K be a field, A an associative unitary K-algebra and $a \in Sep(A)$. The K-algebra $K[a]$ is finite-dimensional, associative and unitary. Based on corollary 8 exactly one pair $(r; v) \in rad(K[a]) \times VSep(K[a])$ exists such that $a = r + v$ is valid. We want to construct this pair. Let $f_1^{k_1} \cdots f_n^{k_n}$ be the decomposition into irreducible polynomials of $min_{a,K}$ in $K[t]$, for all $i \in \underline{n}$ let $B_i := K[t]/(f_i^{k_i})$, $B := \bigoplus_{i=1}^{n} B_i$ and $C := K[t]/(min_{a,K})$. In addition, let F_a be the well-known $\mathcal{A}_1$-isomorphism from $K[a]$ to C (the inverse map of $f + (min_{a,K}) \mapsto f(a)$) and χ the $\mathcal{A}_1$-isomorphism between C and B from the Chinese Remainder theorem. The following diagram is useful for the calculation done afterwards:

$$\begin{array}{ccc}
K[a] & \xrightarrow{F_a \chi} & B \\
\uparrow {+_C F_a^{-1}} & & \downarrow {(Z_{B_1},...,Z_{B_n})} \\
C \times C & \xleftarrow{S(B_i,n)\,(\chi^{-1},\chi^{-1})} & \bigoplus_{i=1}^{n} B_i \times B_i
\end{array}$$

We calculate $aF_a\chi = (t + (f_1^{k_1}), \cdots, t + (f_n^{k_n}))$. The local situation in proposition 12 is used to decompose for all $i \in \underline{n}$ the element $t + (f_i^{k_i})$. Thus, for all $i \in \underline{n}$ we choose elements $r_i, s_i \in K[t]$ such that $t + (f_i^{k_i}) = (r_i + (f_i^{k_i})) + (s_i + (f_i^{k_i}))$, $r_i + (f_i^{k_i}) \in rad(B_i)$ and $s_i + (f_i^{k_i}) \in VSep(B_i)$ are true. We deduce that a is identical to

$$(r_1+(f_1^{k_1}),\cdots,r_n+(f_n^{k_n}))\chi^{-1}F_a^{-1} + (s_1+(f_1^{k_1}),\cdots,s_n+(f_n^{k_n}))\chi^{-1}F_a^{-1}.$$

Within this presentation the first summand is nilpotent and the second one is fully separable in $K[a]$. Later on we will calculate some examples for this decomposition.◇

Definition 9 *(generalized Jordan decomposition)* Let K be a field, A an unitary associative K-algebra and $a \in A$. A pair $(r; s) \in A \times A$ is called a generalized Jordan decomposition of a in A if $a = r+s$, $rs = sr$, $r \in Nil(A)$ and $s \in VSep(A)$ are valid. If $a \in Sep(A)$ is valid, then theorem 37 can be used to construct a generalized Jordan decomposition for a.

We remark that within chapter 1.4 of [50] a Jordan decomposition is defined for linear functions possessing a separable minimal polynomial. This decomposition is relevant within the theory of Lie algebras.◇

In the next remark we present a dual approach to the top-down calculation within chapter 2:

Remark 28 *(linking radical complements and Jordan decomposition for solvable algebras: bottom-up calculation)* Let K be a field, A a finite-dimensional associative unitary K-algebra, $a \in A$ and $A/rad(A)$ separable.

(i) If A is commutative, then theorem 36 is used to prove

$$A = Sep(A) = VSep(A) \oplus_K rad(A).$$

Based on theorem 37 the Jordan decomposition of a and the representation of a based on $rad(A)$ and $VSep(A)$ are identical.

(ii) In general the Jordan decomposition of a and the representation of a based on the nilradical and a radical complement are not identical. If A is solvable, then theorem 26 is used to deduce $rad(K[a]) = rad(A) \cap K[a]$. Theorem 26 ensures the existence of a radical complement T in A such that $T \cap K[a]$ is a radical complement in $K[a]$. Based on T both decompositions are identical.

This argument can be enhanced to so-called basic or reduced algebras. Basic associative algebras A are characterized by the condition $nil(A) = rad(A)$. It is straightforward to prove that this property is equivalent to the fact

Commutative algebras   165

that $A/rad(A)$ is a direct sum of division algebras. As a consequence, every subalgebra of a basic algebra is basic, and thus for every subalgebra T of a basic algebra A the condition $rad(T) = rad(A) \cap T$ is valid. (This may be proven by the reader as an exercise. Alternatively, see [51] or proposition 15.) If $a = r + s$ is a Jordan composition of a within a basic algebra A, then we consider the subalgebra $K[a]$. Based on theorem 38 and corollary 8 we deduce that $K[a]$ possesses a radical complement X, and the nilradical is contained in the nilradical of A (because A is basic). By using corollary 4 we can extend X to a radical complement T of A. The decomposition of a within $rad(A)$ and T as well as the Jordan decomposition are identical.

It seems likely to ask for which algebras the following condition is valid: For every separable element a possessing a Jordan decomposition $a = r + t$ exists a radical complement T of A such that the Jordan decomposition is identical to the decomposition of a within T and $rad(A)$. We prove that $nil(A) = rad(A)$ is true, and thus A is basic. Let a a nilpotent element of A. Then $a = a + 0$ is the Jordan decomposition of a. Let T be a radical complement of A such that the Jordan decomposition is identical to the decomposition of a within T and $rad(A)$. We deduce $a \in rad(A)$.

As a consequence for every separable element within a basic algebra a radical complement T exists such that a can be decomposed as $a = r+t \in rad(A) \oplus T$ and $r \circ t = 0$ is valid (because the Jordan decomposition has this property). Thus, based on the Jordan decomposition have calculated for every separable element within a basic algebra also a decomposition of the element linked to a decomposition of the algebra into the radical and a radical complement without calculating explicitly a radical complement. We call this method bottom-up calculation in contrast to the top-down calculation presented within section 2.7. Within part (iv) we will use this to describe a radical complement also bottom-up.

(iii) For part (ii) an example is presented now. Let $K := GF(3)$, $A := \Delta_{u,3}$ and $a := \begin{pmatrix} 2_K & 0_K & 0_K \\ 1_K & 1_K & 0_K \\ 2_K & 1_K & 2_K \end{pmatrix}$ (see section 1.3.2). Within A we calculate

$a = \begin{pmatrix} 0_K & 0_K & 0_K \\ 1_K & 0_K & 0_K \\ 2_K & 1_K & 0_K \end{pmatrix} + \begin{pmatrix} 2_K & 0_K & 0_K \\ 0_K & 1_K & 0_K \\ 0_K & 0_K & 2_K \end{pmatrix}$. The first resp. second summand is contained in $rad(A)$ resp. in $D(3,K)$. In addition, $min_{a,K} = (t + 1_K)(t - 1_K)$ is valid. Hence, $a \in VSep(A)$ is true. This examples shows that both decompositions are not identical in general. A is solvable and a is fully separable. Thus, $K[a]$ is separable, too (see corollary 8). We use part (i) of theorem 4 and remark 10 to prove the existence of an element $r \in rad(A)$ such that $K[a]^{1_A + r} \subseteq D(3,K)$ and $K[a] \subseteq D(3,K)^{(1_A + r)^{-1}}$ are valid. Within $K[a]^{1+r}$ both decompositions are identical. Within this special example the element r can be determined by the recursion algorithm 1: suitable conjugators are of the form $\begin{pmatrix} 1_K & 0_K & 0_K \\ 1_K & 1_K & 0_K \\ b & 2_K & 1_K \end{pmatrix}$ with arbitrary $b \in K$.

(iv) We use some parts of the top-down calculation within chapter 2.7 to describe a radical complement 'bottom-up'. For a solvable associative algebra with separable radical complement we have proven within corollary 10 that every element of A is separable. If we assume that A is basic, then this is not clear in general. For our argumentation we assume that every element is separable (e.g. that K is perfect). Let B be a basis of A. Because of the separability of each element we can construct a Jordan decomposition $b = r_b + t_b$ for every $b \in B$ such that $r_b \in rad(A)$ and t_b is fully-separable are valid. Here we use that for basic algebras the set of nilpotent elements is exactly the nilradical (see proposition 15). By the generalized conjugacy part of the Wedderburn-Malcev theorem (see corollary 4) for every $b \in B$ we can extend the fully-separable element t_b (because $K[t_b]$ is a separable subalgebra, see corollary 8) to a radical complement T_b. Let us fix one explicit basic element c. Then for every $b \in B \setminus \{c\}$ we use again corollary 4 to derive conjugators $x_b \in rad(A)$ such that

$$T_c = T_b^{1+x_b}$$

is valid. Now we can apply the transfer rule from chapter 2.7:

$$b = r_b + t_b = \hat{r}_b + t_b^{1+x}$$

such that

$$\hat{r}_b = r_b - t_b - t_b^{1+x_b} \in rad(A) \text{ and } t_b^{1+x_b} \in T_c = T_b^{1+x_b}$$

are valid. Now we derive

$$A = \langle B \rangle_K \subseteq \langle \hat{r}_b, r_c, c \neq b \in B \rangle_K \oplus \langle t_b^{1+x_b}, t_c, c \neq b \in B \rangle_K \subseteq rad(A) \oplus T_c.$$

Comparing dimensions we deduce

$$rad(A) = \langle \hat{r}_b, r_c, c \neq b \in B \rangle_K \text{ and } T_c = \langle t_b^{1+x_b}, t_c, c \neq b \in B \rangle_K.$$

The problem for a calculation is to find these conjugators x_b. All other steps can be calculated explicitly. Nevertheless, for a description of one radical complement for basic algebras this argumentation is valid. It is highly connected to the fact that $rad(A) = nil(A)$ is true.⋄

Commutative algebras

5.4.2 Properties of the generalized Jordan decomposition

For the next theorem linked to properties of the generalized Jordan decomposition we need the following remark about eigenvalues. Within part (iii),(d) of the theorem we link the generalized Jordan decomposition to the Jordan decomposition for splitting endomorphism.

Remark 29 Let K be a field, $f, g \in K[t]$, V a finite-dimensional K-space, $\alpha \in End_K(V)$, $\alpha = f(\alpha) + g(\alpha)$ and $f(\alpha)$ nilpotent. For every eigenvalue k of α the condition $k = g(k)$ is valid.

Proof. Let k be an eigenvalue of α with corresponding eigenvector v. $f(k)$ resp. $g(k)$ is an eigenvalue of $f(\alpha)$ resp. of $g(\alpha)$ with eigenvector v. The nilpotency of $f(\alpha)$ lets us deduce that $f(k) = 0_K$ is valid. Hence,

$$kv = v\alpha = vf(\alpha) + vg(\alpha) = g(k)v$$

is true. We use $v \neq 0_V$ to finish the proof.⋄

Theorem 38 *(properties of the generalized Jordan decomposition)* Let K be a field, A an associative unitary K-algebra and $a \in A$. The following statements are valid:

(i) The following statements are equivalent:

 (a) $a \in Sep(A)$
 (b) a possesses a generalized Jordan decomposition in $K[a]$.
 (c) a possesses a generalized Jordan decomposition in A.

(ii) a possesses at least one generalized Jordan decomposition in A.

(iii) Let $(r; s)$ be a generalized Jordan decomposition of a in A and $f, g \in K[t]$ such that $r = f(a)$ and $s = g(a)$ are valid. The following statements are valid:

 (a) $min_{s,K} = halb(min_{a,K})$
 (b) $cl(r) = max(min_{a,K})$, $min_{r,K} = t^{cl(r)}$
 (c) Let $p, q \in K[t]$. $(p(a); q(a))$ is a generalized Jordan decomposition of a in A if and only if $p \in f + (min_{a,K})$ and $q \in g + (min_{a,K})$ are valid.
 (d) $a \in ZF(A)$ is valid if and only if $r \in D(A)$ is true.
 (connection to the Jordan decomposition of splitting endomorphism)

Proof. ad(i): The implication from (a) to (b) is contained in theorem 37. In addition, the implication from (b) to (c) is straightforward to prove. Let $(r;s)$ be a generalized Jordan decomposition of a in A. We define $T := K[r,s]$. Because of $rs = sr$ the K-space T is a commutative associative unitary K-algebra. r, s are algebraical over K, and thus the K-algebras $K[r]$ and $K[s]$ are finite-dimensional and centralizing each other. Hence, T is finite-dimensional as homomorphic image of $K[r] \otimes_K K[s]$. Straightforward to prove are that $r \in rad(T)$ and $s \in VSep(T)$ are valid. Based on part (i) of theorem 36 we deduce $a = r + s \in Sep(T) \subseteq Sep(A)$.

ad(ii): Let $(r_1;s_1)$ be a generalized Jordan decomposition of a in A. Based on part (i) the element a possesses a generalized Jordan decomposition $(r;s)$ in $K[a]$, too. Hence, elements $f,g \in K[t]$ exist such that $r = f(a)$ and $s = g(a)$ are valid. We define $T := K[a, r_1]$. a, r_1 are algebraical over K, and thus T is finite-dimensional as homomorphic image of the K-algebra $K[a] \otimes_K K[r_1]$. Finally, we use part (i) of theorem 36 to deduce $Sep(T) = VSep(T) \oplus_K rad(T)$, and part (ii) is proven.

ad(iii), (a): Let $g_1 \cdots g_l$ resp. $f_1^{t_1}, \cdots f_m^{t_m}$ be the decomposition of $min_{s,K}$ resp. of $min_{a,K}$ in $K[t]$ into irreducible polynomials.

(1) At first we prove that $grad(halb(min_{a,K})) = grad(min_{s,K})$ is valid. Let $T := K[a]$. $\langle r \rangle_A \subseteq rad(T)$, $K[s] \subseteq VSep(T)$ (see theorem 35) and $T = K[s] \oplus_K \langle r \rangle_A$ are valid. a is based on part (i) separable over K. We use part (i) of theorem 36 to deduce $T = Sep(T) = rad(T) \oplus_K VSep(T)$. A dimension argument is used to prove $VSep(T) = K[s]$ and $rad(T) = \langle r \rangle_A$. The Chinese Remainder theorem leads to $T/rad(T) \cong_{A_1} \bigoplus_{i=1}^{m} K[t]/(f_i)$ and $K[s] \cong_{A_1} \bigoplus_{i=1}^{l} K[t]/(g_i)$. Because of $T/rad(T) \cong_{A_1} K[s]$ and the main theorem of Wedderburn-Artin about the structure of associative algebras both degrees are identical.

Both polynomials possess the leading coefficient 1, and thus we have only to prove that $halb(min_{a,K}) \mid min_{s,K}$ is valid. This proof is stated in part (2).

(2) Let L be a splitting field of $min_{a,K}$ over K and $B := K[a] \otimes_K L$. B is a finite-dimensional L-space. Let ρ be the right regular representation of $K[a]$. We calculate $a\rho \otimes id_L = f(a\rho \otimes id_L) + g(a\rho \otimes id_L)$. 65.5 in [37] is used to prove that the first summand is nilpotent, and by the choice of L the polynomial $min_{a\rho \otimes id_L, L}$ splits over L. Let l be an eigenvalue of $a\rho \otimes id_L$ in L. Remark 29 is used to deduce $g(l) = l$. a is separable over K, and thus $halb(min_{a\rho \otimes id_L, L}) \mid_L min_{g(a\rho \otimes id_L), L} = min_{s\rho \otimes id_L, L}$ is valid. Again by using 65.5 in [37] (Minimal polynomials are not changing

Commutative algebras   169

under base field extensions.) we conclude that $halb(min_{a\rho,K})\mid_L min_{s\rho,K}$ and $halb(min_{a,K})\mid_L min_{s,K}$ are true. Both polynomials are contained in $K[t]$, and thus part (a) is proven.

ad(b): Let $f_1{}^{t_1}\cdots f_m{}^{t_m}$ be the decomposition into irreducible polynomials of $min_{a,K}$ over K. As proven in part (a) the statements $Sep(K[a]) = K[a]$, $rad(K[a] = \langle r \rangle_A$ and $VSep(K[a]) = K[s]$ are valid. Let $T := K[t]/(min_{a,K})$. $K[a] \cong_{A_1} T$ and $rad(T) = (f_1\cdots f_m)/(min_{a,K})$ are true. In particular, $cl(rad(T)) = cl(rad(K[a]))$ is valid. Let $c := max\{r_1,\cdots,r_m\}$. First we prove $cl(rad(T)) = c$. Because of $min_{a,K} \mid (f_1\cdots f_m)^c$ we deduce $cl(rad(T)) \leq c$. Let us assume $cl(rad(T)) \leq c - 1$ and w.l.o.g. let $cl(rad(T)) \leq r_1 - 1$ be valid. $min_{a,K} \mid (f_1\cdots f_m)^c$ is valid and from this statement we would deduce a contradiction to the decomposition into irreducible polynomials of $min_{a,K}$ over K.
We prove now $cl(r) = cl(rad(K[a]))$. Because of $r \in rad(K[a])$ we deduce $cl(rad(K[a])) \leq cl(r)$. In addition, $rad(K[a]) = \langle r \rangle_A = \{h(r) \mid h \in tK[t]\}$ is true. This is used to finish the proof of part (b).

ad(c): Let $(p(a); q(a))$ be another generalized Jordan decomposition of a in A. Part (ii) let us deduce that $f(a) = p(a)$ and $g(a) = q(a)$ are valid. Thus, $(f-p)(a) = 0_A = (g-q)(a)$ is true. This statement implies $p \in f + (min_{a,K})$ and $q \in g + (min_{a,K})$. If $p \in f + (min_{a,K})$ and $q \in g + (min_{a,K})$ are valid, then $p(a) = f(a)$ and $q(a) = g(a)$ are true.

ad(d): Let $a \in ZF(A) \subseteq Sep(A)$. Based on part (i) the element a possesses a generalized Jordan decomposition $(r; s)$ in A. Part (iii), (a) is used to prove $s \in D(A)$. Let $s \in D(A)$. By using the parts (a) and (b) of statement (iii) we deduce $min_{a,K} \mid min_{s,K}{}^{cl(r)}$. Hence, a is a splitting element.⋄

Before we focus on an example for calculating the generalized Jordan decomposition we pay attention to part (iii) of theorem 38. The result is useful for determining minimal polynomials. This is the content of the next corollary. Within the proof of part (iv) of theorem 38 this method was already used.

Corollary 11 (bounding minimal polynomials) Let K be a field, A an associative unitary K-algebra, $a \in VSep(A)$ and $r \in Nil(A)$. If $ar = ra$ is valid, then $a + r \in Sep(A)$ and $min_{a,K} \mid min_{a+r,K} \mid min_{a,K}{}^{cl(r)}$ are true.

Proof. By definition the pair $(r; a)$ is a generalized Jordan decomposition of $r + a$ in A. Based on parts (i) and (iii) of theorem 38 we finish the proof.⋄

The next example is illustrating this corollary.

Example 13 (i) Let $K := GF(2)(t)$ and $A := A(t, 0_K, K)$ (see example 8). i is a nilpotent element of class 2. We use corollary 11 to deduce $min_{i+1_A,K} \mid$

$(t-1_K)^2$. i is not contained in $K1_A$, and thus $min_{i+1_A,K} = (t-1_K)^2$ is valid.

(ii) Let $K := \mathbb{Q}$, $A := End_K(K^3)$, B the standard basis of K^3 and $\gamma \in A$ defined by

$$A := M_B(\gamma) = \begin{pmatrix} 2 & 0 & 0 \\ 1 & 1 & 0 \\ 1 & 1 & 2 \end{pmatrix}. \text{ Let } N := \begin{pmatrix} 0 & 0 & 0 \\ 0 & 0 & 0 \\ 1 & 1 & 0 \end{pmatrix} \text{ and } V := \begin{pmatrix} 2 & 0 & 0 \\ 1 & 1 & 0 \\ 0 & 0 & 2 \end{pmatrix}.$$

We calculate $A = N + V$, $NV = VN$ and $cl(N) = 2$. $t^2 - 3t + 1$ is irreducible over $\mathbb{Q}$. We use the method of invariant subspaces (see e.g. [37] or the exercises in this work) to obtain $min_{V,K} = (t-2)(t^2 - 3t + 1)$. Based on corollary 11 we deduce $(t-2)(t^2 - 3t + 1) \mid min_{A,K} \mid (t-2)^2(t^2 - 3t + 1)^2$. $grad(min_{A,K}) \leq 3$ is valid, and thus $min_{A,K} = (t-2)(t^2 - 3t + 1)$ is true.⋄

The next remark is needed before we focus on an example for the generalized Jordan decomposition within a linear algebraic context.

Remark 30 (i) Based on the generalized Jordan decomposition we have answered one of our **main topics** in this work for finite-dimensional associative commutative unitary algebras possessing a separable factor algebra by its nilradical: the representation of an element as the sum of a nilpotent and of an element of the unique radical complement in a constructive way. If we focus on a K-basis B of the algebra and decompose its elements by using the generalized Jordan decomposition, like $b = b_{nil} + b_{vsep}$ for all $b \in B$, then a dimension argument leads to the statement that $\langle b_{nil} \mid b \in B \rangle_K$ resp. $\langle b_{vsep} \mid b \in B \rangle_K$ is a K-generating set of $rad(A)$ resp. $VSep(A)$.

(ii) We focus on a linear algebraic application of part (i). Let K be a field, A an associative unitary K-algebra and $a \in Sep(A)$. We want to determine a K-basis for $rad(K[a])$ and for $VSep(K[a])$. For this, let $(a_{nil}; a_{vsep})$ be the generalized Jordan decomposition of a in $K[a]$. $\langle a_{nil} \rangle_A \subseteq rad(K[a])$, $K[a_{vsep}] \subseteq VSep(K[a])$ (see theorem 35) and $K[a] = K[a_{vsep}] \oplus_K \langle a_{nil} \rangle_A$ are true. In addition, a is separable. Based on part (i) of theorem 36 we deduce $K[a] = rad(K[a]) \oplus_K VSep(K[a])$. A dimension argument leads to $VSep(K[a]) = K[a_{vsep}]$ and $rad(K[a]) = \langle a_{nil} \rangle_A$. Part (iii), (a) and part (iii), (b) of theorem 38 are used to prove that $\{(a_{nil})^s \mid s \in \overline{max(min_{a,K}) - 1}\}$ resp. $\{(a_{vsep})^s \mid s \in \{0\} \cup \overline{grad(halb(min_{a,K})) - 1}\}$ is a K-basis of $rad(K[a])$ resp. of $VSep(K[a])$.⋄

Example 14 Let $K := \mathbb{R}$, $A := End_K(K^4)$, B the standard basis of K^4 and $\gamma \in A$ defined by the matrix

$$M_B(\gamma) = \begin{pmatrix} 1 & 0 & 0 & 0 \\ 1 & 1 & 0 & 0 \\ 0 & 0 & -1 & 1 \\ 0 & 0 & -2 & 1 \end{pmatrix}.$$

Commutative algebras

We calculate – based on invariant subspaces – $char_{\gamma,K} = min_{\gamma,K} = (t-1)^2(t^2+1)$. Thus, $halb(min_{\gamma,K}) = (t-1)(t^2+1)$ and $max(min_{\gamma,K}) = 2$ are valid. $char(K) = 0$ is true, and hence $min_{\gamma,K} \in Sep(A)$ is valid. Now we calculate the generalized Jordan decomposition of γ in A. For this, we use the method within theorem 37. $\gamma F_\gamma \chi = (t + ((t-1)^2); t + (t^2+1))$ is valid. For the local components we determine the decomposition as stated within proposition 12. $t = 1(t-1) + 1$ and $t = 0(t^2+1) + t$ are valid. We deduce $\gamma F_\gamma \chi = ((t-1),(0)) + ((1),(t))$. The first summand is nilpotent and the second one is fully separable over K. For determining the image of χ^{-1} we have to solve congruences (Chinese Remainder theorem). Polynomials $f, g, h \in K[t]$ are to be calculated such that
$f \equiv t - 1 \bmod (t-1)^2$, $f \equiv 0 \bmod t^2 + 1$,
$g \equiv 1 \bmod (t-1)^2$, $g \equiv 0 \bmod t^2 + 1$ and
$h \equiv 0 \bmod (t-1)^2$, $h \equiv t \bmod t^2 + 1$ are valid.
The solution is[3]
$f = (t-1)(t^2+1)(-\frac{1}{2}t + 1)$,
$g = (t^2+1)(-\frac{1}{2}t + 1)$ and
$h = \frac{1}{2}t^2(t-1)^2$. Thus, $\gamma = f(\gamma) + (g+h)(\gamma)$ is true. The first summand is nilpotent and the second one is fully separable over K. The representation matrices are $M_B(f(\gamma)) = \begin{pmatrix} 0 & 0 & 0 & 0 \\ 1 & 0 & 0 & 0 \\ 0 & 0 & 0 & 0 \\ 0 & 0 & 0 & 0 \end{pmatrix}$ and

$M_B((g+h)(\gamma)) = \begin{pmatrix} 1 & 0 & 0 & 0 \\ 0 & 1 & 0 & 0 \\ 0 & 0 & -1 & 1 \\ 0 & 0 & -2 & 1 \end{pmatrix}$.

The first matrix is strict lower triangular. In addition, $cl(f(\gamma)) = 2$ and $min_{(g+h)(\gamma),K} = (t-1)(t^2+1)$ are valid.
Part (ii) of remark 30 is used to deduce that $\{f(\gamma)\}$ resp. $\{1_A, (g+h)(\gamma), ((g+h)(\gamma))^2\}$ is a K-basis of $rad(K[\gamma])$ resp. of $VSep(K[\gamma])$.◊

If the fully separable part is diagonalizable, then – by Jordan's theorem – a basis of the K-space exists such that the representation matrix linked to this basis of the diagonalizable resp. nilpotent part is a diagonal resp. strict lower triangular matrix. A similar result is valid in general, and again we use the method within corollary 11 within the proof.

Corollary 7 *Let K be a field, V a finite-dimensional K-space, $\varphi \in Sep(End_K(V))$ and $(r; s)$ a generalized Jordan decomposition of φ in $K[\varphi]$. The following statements are valid:*

---
[3]This calculation is included in standard courses of linear algebra I and II. We will focus on this determination within the exercises.

(i) The summands of the primary decomposition of V into $K[\varphi]$-modules are $K[s]$-isomorphic to the irreducible $K[s]$-modules of the $K[s]$-modules V.

(ii) For every summand W of the primary decomposition of V into $K[\varphi]$-modules a basis B_W of W exists such that $M_B(r|_W)$ is a strict lower triangular matrix.

Proof. ad(i): Let $g \in K[t]$ such that $s = g(\varphi)$ is true. All summand are s-invariant. By using part (iii),(a) of theorem 38 and the theorem of Krull-Remak-Schmidt we deduce part (i).

ad(ii): Because of $r \in K[\varphi]$ all summands are r-invariant, and thus we deduce part (ii).⋄

5.5 The subalgebras of splitting and diagonizable elements

Within part (iii),(d) of theorem 38 the sets $D(A)$ and $ZF(A)$ are of special interest. They are linking the generalized Jordan decomposition to the well-known Jordan decomposition of splitting endomorphism. We start the analyze with the set of diagonalizable elements.

Theorem 39 *(the subalgebra of diagonalizable elements)* Let K be a field and A an associative commutative unitary K-algebra. $D(A)$ is an unital K-subalgebra of A. If $D(A)$ is finite-dimensional, then $D(A)$ is separable and $D(A) \cong_{A_1} K^{dim_K(D(A))}$ is valid.

Proof. It is straightforward to prove that $K1_A \subseteq D(A)$ is valid. Let $a, b \in D(A)$. By using the Chinese-Remainder theorem elements $r, s \in \mathbb{N}$ exists such that $K[a] \cong_{A_1} K^r$ and $K[b] \cong_{A_1} K^s$ are true. Thus, $K[a] \otimes_K K[b] \cong_{A_1} K^{rs}$ is valid. In addition, $K[a,b]$ is generated by the commutating subalgebras $K[a]$ and $K[b]$. Therefor, it is a homomorphic image of $K[a] \otimes_K K[b] \cong_{A_1} K^{rs}$. Based on part (iii) of theorem 26 for every ideal I of K^{rs} an element $t \in \mathbb{N}$ exists such that $I \cong_{A_1} K^t$ is valid. K^{rs} is semisimple, and thus I possesses an ideal complement in K^{rs}. We deduce the existence of an element $w \in \mathbb{N}$ such that $K^{rs}/I \cong_{A_1} K^w$ is true. Hence, an element $p \in \mathbb{N}$ exists possessing the property $K[a,b] \cong_{A_1} K^p$. This argumentation is also valid for proving the add-on: instead of elements a, b a finite basis of $D(A)$ is to be used. If $x \in K[a,b]$ is true, then – because of part (iii) of theorem 26 – an element $d \in \mathbb{N}$ exists such that $K[x] \cong_{A_1} K^d$ is valid. The Chinese-Remainder theorem lets us conclude $x \in D(A)$.⋄

Theorem 40 *(connection to the subalgebra of splitting elements)* Let K be a field and A a finite-dimensional associative commutative unitary K-algebra. The following statements are valid:

Commutative algebras   173

(i) $ZF(A) = rad(A) \oplus_K D(A)$ is valid. In particular, $ZF(A)$ is a K-subalgebra of A possessing the unique algebra complement $D(A)$ of $rad(ZF(A)) = rad(A)$ in $ZF(A)$.

(ii) $D(A)$ is the unique maximal semisimple K-subalgebra of A such that K is a splitting field. In particular, $A = D(A)$ is valid if and only if A is $\mathcal{A}_1$-isomorphic to $K^{dim_K(A)}$.

(iii) $ZF(A)$ is the unique maximal unital K-subalgebra of A for which K is a splitting field.

(iv) If $A/rad(A)$ is separable, then $H(A) = VSep(A)$ is valid. In particular, under this assumption $H(A)$ is a K-subalgebra of A.

(v) In general, $H(A)$ is no K-subalgebra of A.

Proof. ad(i): This part is deductable by the statements (i) and (iii),(d) of theorem 38.

ad(ii): Based on theorem 39 the K-subalgebra $D(A)$ possesses the desired properties. Let T be a K-subalgebra of A such that an element $r \in \mathbb{N}$ exists possessing the property $T \cong_{\mathcal{A}_1} K^r$. Let $x \in T$. Because of part (iii) of theorem 26 another element $s \in \mathbb{N}$ exists such that $K[x] \cong_{\mathcal{A}_1} K^s$ is valid. By using the Chinese-Remainder theorem we conclude $x \in D(T) \subseteq D(A)$.

ad(iii): Based on parts (i) and (ii) the set $ZF(A)$ possesses the desired properties. Let T be an unital K-subalgebra of A such that an element $r \in \mathbb{N}$ exists possessing the property $T/rad(T) \cong_{\mathcal{A}_1} K^r$. $T/rad(T)$ is separable, and thus based on theorem 13 an algebra complement D of $rad(T) = rad(A) \cap T$ in T exists. Part (ii) lets us deduce that $D \subseteq D(A)$ is valid. Hence, $T \subseteq rad(A) + D(A)$ is valid, and because of part (i) the statement $T \subseteq ZF(A)$ is true.

ad(iv): Let $A/rad(A)$ be separable. $VSep(A) \subseteq H(A)$ is true in general. Let $x \in H(A)$. Based on the Chinese-Remainder theorem we deduce that $K[x]$ is a semisimple K-subalgebra of A. A is solvable, and thus – based on part (ii) of theorem 26 – the semisimple K-subalgebra $K[x]$ is separable. Part (ii) of lemma 7 lets us deduce that x in $VSep(A)$ is valid.

ad(v): Let $char(K) = 2$, $K \setminus QA(K) \neq \emptyset$ and $a \in K \setminus QA(K)$. We focus on the K-algebra $A := A(a, 0_K, K)$. Because of example 8 we know that $rad(A) = \langle j, k \rangle_K$ is valid and that $\langle 1_A, i \rangle_K$ is a algebra complement of $rad(A)$ in A. In addition, we calculate $i^2 = (i+j)^2 = a1_A$. Hence, $min_{i,K} = min_{i+j,K} = t^2 + a$ is valid and $t^2 + a$ is irreducible over $K[t]$. Therefor, $i, i+j \in H(A)$ is true. Because of $j \in rad(A)$ the set $H(A)$ is no

K-subspace.◇

The following Hasse diagrams[4] illustrate the proven results:

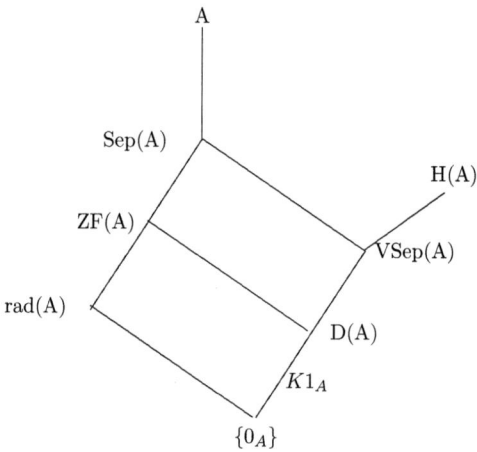

Parts (ii)-(iv) of example 11 can be illustrated by the following Hasse diagrams:

---

[4]Helmut Hasse (German; born 25th August 1898; died 26th December 1979) was a German mathematician working in algebraic number theory, known for fundamental contributions to class field theory, the application of p-adic numbers to local class field theory and diophantine geometry (Hasse principle), and to local zeta functions. Hasse was born in Kassel, Province of Hesse-Nassau, the son of Judge Paul Reinhard Hasse aka Haße (12nd April 1868 to 1st June 1940, son of Friedrich Ernst Hasse and his wife Anna Von Reinhard) and his wife Margarethe Louise Adolphine Quentin (born 5th July 1872 in Milwaukee, daughter of retail toy merchant Adolph Quentin (b. May 1832, probably Berlin, Kingdom of Prussia) and Margarethe Wehr (about 1840, Prussia), then raised in Kassel). After serving in the Imperial German Navy in World War I, he studied at the University of Göttingen, and then at the University of Marburg under Kurt Hensel, writing a dissertation in 1921 containing the Hasse-Minkowski theorem (as it is now called) on quadratic forms over number fields. He then held positions at Kiel, Halle and Marburg. He was Hermann Weyls replacement at Göttingen in 1934. Hasse was an Invited Speaker of the ICM in 1932 in Zurich and a Plenary Speaker of the ICM in 1936 in Oslo. In 1933 Hasse had signed the Loyalty Oath of German Professors to Adolf Hitler and the National Socialist State. Politically, he applied for membership in the Nazi Party in 1937, but this was denied to him due to his Jewish ancestry. After the war, he briefly returned to Göttingen in 1945, but was excluded by the British authorities. After brief appointments in Berlin, from 1948 on he settled permanently as professor in Hamburg. He collaborated with many mathematicians, in particular with Emmy Noether and Richard Brauer on simple algebras, and with Harold Davenport on Gauss sums (Hasse-Davenport relations), and with Cahit Arf on the Hasse-Arf theorem.

(ii) $A := A(0_K, 0_K, K)$, $char(K) = 2$

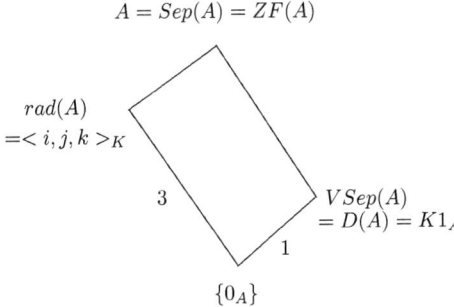

(iii) $A := A(a, 0_K, K)$, $char(K) = 2$, $a \notin QA(K)$

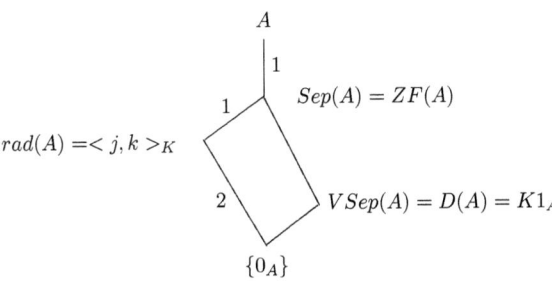

(iv) $A := A(a, b, K)$, $char(K) = 2$, A is a field

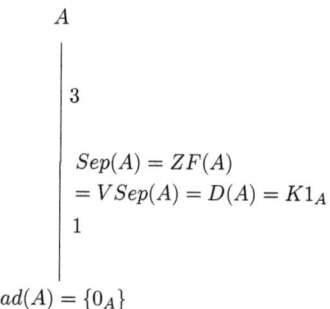

Before we transfer the results of the last four sections to non-unitary commutative algebras we apply the theory to commutative group algebras.

5.6 Commutative group algebras

Within this section let G be a finite Abelian group and K a field.
We analyze the group algebra KG within the semisimple and modular case. This differentiation is based on the well-known theorem of Maschke: the group algebra is semisimple if and only if the characteristic of K does not divide the order of G. Otherwise the group algebra is called modular. The aim of this section is to determine the K-subalgebras $H(KG)$, $D(KG)$, $VSep(KG)$, $ZF(KG)$, $rad(KG)$ and $Sep(KG)$. In addition, we will calculate the generalized Jordan decomposition for the elements of $Sep(KG)$. By aug we denote the augmentation function from KG onto K. Its kernel is the augmentations ideal $Aug(KG)$.

Proposition 13 *(semisimple Abelian group algebras) Let KG be semisimple. The following statements are valid:*

(i) $rad(KG) = 0$

(ii) $H(KG) = Sep(KG) = VSep(KG) = KG$

(iii) $ZF(KG) = D(KG)$

(iv) *For all $a \in KG$ the pair $(0; a)$ is the generalized Jordan decomposition of a.*

Proof. Because of the semisimplicity of KG the statement $rad(KG) = \{0_{KG}\}$ is true. Based on theorem 2 the group algebra KG is separable. Theorem 36 lets us deduce that $KG = Sep(KG) = VSep(KG)$ is valid. In

Commutative algebras 177

particular, part (iv) is true. We finish the proof by using theorem 40 for deducing the identity $ZF(KG) = D(KG)$. ⋄

The question of determining $D(KG)$ is to be solved. The answer will be provided at the end of this section because it is also the central topic to be analyzed within the modular case.

Proposition 14 *(modular Abelian group algebras) Let KG be modular, p a prime number, $char(K) = p$, S_p the normal p-Sylow subgroup and H the normal complement of S_p in G. The following statements are valid:*

(i) $rad(KG) = KG\,Aug(KS_p) = Aug(KS_p)KG$

(ii) KH is the unique radical complement which is separable.

(iii) $H(KG) = VSep(KG) = KH$

(iv) $KG = Sep(KG)$
    In particular, every element of KG possesses a generalized Jordan decomposition.

(v) $ZF(KG) = rad(KG) \oplus D(KG)$

(vi) $D(KG) = D(KH)$

(vii) If for every element $g \in G$ the pair $(g_{nil}; g_{vsep})$ is a generalized Jordan decomposition, then $(\sum_{g \in G} k_g g_{nil}; \sum_{g \in G} k_g g_{vsep})$ is a generalized Jordan decomposition for $\sum_{g \in G} k_g g$.

(viii) If $s \in S_p$, $h \in H$ and $(s_{nil}; s_{vsep})$ is a generalized Jordan decomposition of s, then $(s_{nil}h; s_{vsep}h)$ is a generalized Jordan decomposition of sh.

(ix) For every element $s \in S$ the pair $((s - aug(s))1_G; aug(s)1_G)$ is a generalized Jordan decomposition of s.

Proof. KG is not semisimple, and thus a prime number p exists such that $|G|$ is divided by it and $char(K) = p$ is valid. Let S_p be the unique p-Sylow subgroup of G and H the normal complement of S_p in G. The proposition in section 4.7 in the text book [35] lets us deduce that $rad(KG) = KG\,Aug(KS_p)$ is valid and that the factor algebra by the nilradical of KG is isomorphic to KH. Based on proposition 13 the subalgebra KH is separable. Hence, theorem 36 lets us deduce that the identities $KG = Sep(KG)$ and $KH = VSep(KG)$ are true. For determining $ZF(KG)$ we apply theorem 40 which reduces the determination to the description of $D(KG)$. Because of theorem 40 this subalgebra is identical to $D(KH)$. The statements for the generalized Jordan decompositions are straightforward to verify.◊

Again, we need to determine the subalgebra of diagonizable elements for a commutative semisimple group algebra. A basis is provided within the next result.

Theorem 41 *(basis for $D(A)$) Let A be a finite-dimensional associative semisimple commutative K-algebra and I the set of unit elements of all minimal ideals of A. I is a K-basis of $D(A)$. In particular, $D(A)$ is isomorphic to K^n for n being the number of minimal ideals of A.*

Proof. It is well-known that I is a linear independent set of A. Based on theorem 39 an element $r \in \mathbb{N}$ exists such that $D(A)$ is isomorphic to K^r. Hence, $D(A)$ possesses a basis containing only idempotents of A. Every idempotent of A is contained in $\langle I \rangle_K$. Thus, $D(A)$ is contained in this set, too. Because of theorem 39 the set $D(A)$ is a K-subalgebra of A. Therefor, we need only to prove that every idempotent of A is diagonalizable. Let e be an idempotent of A. Because of $e^2 = e$ we deduce $min_{e,K}$ is dividing $t(t - 1_K)$.◊

For a semisimple group algebra the set I can be determined by using character theory. The corresponding result is included in [28], chapter 7. (This chapter may be the basis for the reader to get a first insight into character and representation theory of finite groups and algebras.)

Theorem 42 *(irreducible characters and primitive idempotents) Let G be a finite (not necessarily Abelian) group and K be a field such that KG is semisimple. Finite many irreducible characters $\chi_1, \cdots, \chi_h$, exist such that the elements*

$$e_i := \frac{\chi_i(1)}{|G|} \sum_{g \in G} \chi_i(g^{-1}) g$$

are the unit elements of the minimal ideals of KG which decompose KG direct.◊

For applying theorem 42 we need to determine the irreducible characters of Abelian groups. S. Perlis and G.L. Walker present in [34] a decomposition of semisimple commutative group algebras which is used for this purpose and for determining – based on theorem 41 – the dimension of the subalgebra of diagonalizable elements as well as the number and degrees of all irreducible characters. Let r be the order of G, for every divisor d of r the element ω_d a primitive d-th root of unity (in a sufficient field extension of K), $d_d := dim_K(K(\omega_d))$, $o_d := |\{a \in G \mid o(a) = d\}|$ and $a_d := \frac{o_d}{d_d}$. The following theorem is valid:

Theorem 43 *(Perlis/Walker: decomposition of commutative semisimple group algebras)* Let KG be semisimple and r be the order of G. KG is isomorphic to $\bigoplus_{d|r} K(\omega_d)^{a_d}$. For every divisor d of r exactly a_d irreducible characters of degree $dim_K(K(\omega_d))$ do exist. In particular, $D(KG)$ is of dimension $\sum_{d|r} a_d$. ⋄

The subalgebras $D(KG)$ varies for different values of the field K. Within the following corollary this is presented for two fields. Let φ be the Euler's totient function:

Corollary 12 Let r be the order of G. The following statements are valid:

(i) $\mathbb{C}G$ is isomorphic to $\mathbb{C}^r$. In particular, $\mathbb{C}G$ is diagonalizable. $|G|$ different irreducible characters of degree 1 exist.

(ii) $\mathbb{Q}G$ is isomorphic to $\bigoplus_{d|r} \mathbb{Q}(\omega_d)^{a_d}$. In particular, $D(\mathbb{Q}G)$ is of dimension $\sum_{d|r} \frac{|\{a \in G | o(a)=d\}|}{\varphi(d)}$. For every divisor d of r there are $\frac{|\{a \in G | o(a)=d\}|}{\varphi(d)}$ different irreducible characters of degree $\varphi(d)$. ⋄

We focus on the following two examples linked to theorem 43:

Example 15 Let us focus on the cyclic group Z_3 of order 3 and the field of real numbers. For the divisor 1 we calculate one time the summand $\mathbb{R}$ within the decomposition of $\mathbb{R}Z_3$. For the divisor 3 we calculate $a_3 = 1$, because there are two elements of order three and the real numbers possess no primitive 3rd root of unity. Thus, the decomposition of $\mathbb{R}Z_3$ is $\mathbb{R} \oplus \mathbb{C}$. Furthermore, the dimension of $D(\mathbb{R}Z_3)$ is 2. Two different irreducible characters do exist. ⋄

Example 16 We focus on the cyclic group Z_4 of order 4 and the rational numbers. Straightforward to calculate is the statement $a_1 = a_2 = a_4 = 1$. Thus, the dimension of $D(\mathbb{R}Z_4)$ is exactly 3. Three irreducible characters are existing, and the group algebra $\mathbb{Q}Z_4$ is decomposed as $\mathbb{Q} \oplus \mathbb{Q} \oplus \mathbb{Q}(i)$. ⋄

We finalize this section by presenting a construction of the primitive idempotents of semisimple commutative group algebras. The primitive idempotents are the unit elements of its minimal ideals. This construction is not using the irreducible characters explicitly. Based on this construction the subalgebra $D(KG)$ can be determined explicitly, its dimension is already calculated. (The irreducible modules are known for commutative semisimple group algebra. This topic is presented within the exercises.) I want to say thank you to Prof. Geoffrey Robinson for his helpful advices concerning this topic (see [62]).

Construction 2 Let G be of order r and KG be semisimple. $char(K)$ is no divisor of r. First we focus on the special case that K is containing all r-th roots of unity. In this case K is a so-called splitting field. We determine the irreducible characters. If G is cyclic and generated by an element g, then the irreducible characters are exactly the r different homomorphism of G into $E(K)$ such that g is mapped on one of the r-th root of unity. If G is an Abelian finite group, then G is the direct product of cyclic groups like $G_1 \times \cdots \times G_l$. For each factor G_i of this decomposition we can determine its personal irreducible characters as presented for cyclic groups like $\chi_{i,1}, \cdots, \chi_{i,s_i}$. All irreducible characters of G are products of length l like $\chi_{1,j_1} \cdots \chi_{l,j_l}$ such that $j_i \in \underline{s_{ij}}$ is valid. The product of functions is defined by multiplication of the function values. In this case it is also possible to determine the primitive idempotents as described within theorem 42.

Now we focus on the field extension $(K; K(\omega_r))$ such that ω_r is a primitive r-th root of unity. $L := K(\omega_r)$ is a splitting field for G. By the first part of this section we know how to determine the irreducible characters of LG. Let $\chi_1, \cdots, \chi_h$ be these irreducible characters. $char(K)$ does not divide r, and thus the polynomial $t^r - 1$ is separable and $(K; L)$ is a Galois extension. The Galois group $Gal(L; K) = Aut_K(L)$ acts on the set of irreducible characters of LG. This set is $Irr_L(G) := \{\chi_1, \cdots, \chi_h\}$. The action is the composition of functions. If $\chi \in Irr_L(G)$ and $\sigma \in Gal(L; K)$, then we define $\omega := \chi \sigma$ and the element

$$e_\omega := \frac{1}{|G|} \sum_{g \in G} ((\chi_i(g^{-1})\sigma)g.$$

$Irr_L(G)$ is decomposed by this group actions into orbits, like $\Omega_1, \cdots, \Omega_w$. For every orbit Ω_i we define the element

$$e_i := \frac{1}{|G|} \sum_{\omega \in \Omega_i} e_\omega.$$

All primitive idempotents are created by this construction: they are the e_i's.◊

5.7 Non-unitary commutative associative algebras

Within this section we transfer the results of the sections 2 to 5 to commutative and non-unitary associative algebras. For this, we use the following general assumptions within this section:

Within this section let K be a field and A an associative K-algebra.

The main topic is to generalize the definition of algebraical, separable etc. for non-unitary algebras. This is realized within definition 10 and theorem 44. Again, we use the concept of the adjunction of an unit.

Commutative algebras 181

Definition 10 Let $a \in A$. We call a algebraical, splittable, diagonalizable, separable, fully separable resp. semisimple over K if a possesses this characteristic within the K-algebra A^K. If a is algebraical over K, then let $\widetilde{min}_{a,K}$ be the minimal polynomial over K of a as element of A^K.⋄

Now we analyze how to define the minimal polynomial without using the algebra A^K.

Definition and remark 10 Let $a \in A$. We define $\widetilde{F}_a : tK[t] \longrightarrow A, f \longmapsto f(a)$. $\widetilde{F}_a$ is an algebra homomorphism possessing the image $Im\widetilde{F}_a = \langle a \rangle_A$. In addition, $\langle a \rangle_A$ is an ideal of $K[a]$ such that $K[a] = \langle a \rangle_A \oplus_K K1_{A^K}$ is valid. Hence, $K[a] \cong_{A_1} (K, \langle a \rangle_A)$ is true.⋄

Lemma 8 Let $a \in A$. The following statements are valid:

(i) a is algebraical over K if and only if $\widetilde{F}_a$ is not injective.

(ii) If a is algebraical over K, then $ker\widetilde{F}_a = \widetilde{min}_{a,K}K[t]$ is valid.

Proof. The opposite implication within (i) is straightforward to prove. Let a algebraical over K. $t \mid \widetilde{min}_{a,K}$ is valid because otherwise we obtain a contradiction to part (vi) of remark 7. Thus, the implication $\Longrightarrow$ in (i) and the inclusion $\supseteq$ in (ii) are true. Let a be algebraical over K and $g \in ker\widetilde{F}_a$ be valid. Elements $u, h \in K[t]$ exists such that $g = u \cdot \widetilde{min}_{a,K} + h$ and $h = 0_K$ or $grad(h) < grad(\widetilde{min}_{a,K})$ are valid. Because of $h(a) = 0_K$ we deduce $h = 0_K$.⋄

We analyze the compatibility of the new definition with the one for unitary algebras. For this, we need the next definition and lemma.

Definition 11 *(set of zero divisors)* By $N(A)$ we denote the set of (right and left) zero divisors of A.⋄

Lemma 9 Let A be finite-dimensional and unitary and T be an unital K-subalgebra of A. The statements $E(T) = E(A) \cap T$, $N(T) = N(A) \cap T$ and $T = N(T) \dot{\cup} E(T)$ are valid.

Proof. $E(T) \subseteq E(A) \cap T$ is valid. Let $x \in E(A) \cap T$. Based on part (2) of theorem 1.2.1 in [8] the set T is a disjoint union of $E(T)$ and $N(T)$. If x would be a zero divisor of T, then x would be a zero divisor of A. This is a contradiction to the choice of x. A similar argumentation is used to prove the second part.⋄

Corollary 13 *(characterization of units and zero divisors)* Let A be finite-dimensional and unitary and $a \in A$. a is an unit resp. zero a divisor of A if and only if t is a divisor resp. no divisor of $min_{a,K}$ ist.

Proof. Based on lemma 9 we only need to prove one equivalence. We use lemma 9 to deduce the following argumentation:
$a \in N(A)$
$\iff a\rho \in N(A\rho)$
$\iff a\rho \in N(End_K(A)) \cap A\rho$
$\iff a\rho \in N(End_K(A))$
$\iff \ker a\rho \neq \{0_A\}$
$\iff 0_K$ is an eigenvalue of $a\rho$
$\iff t \mid min_{a\rho,K}$
$\iff t \mid min_{a,K}.\diamond$

Remark 31 Let A be unitary and $a \in A$. If a is – as element of the algebra A – algebraical over K, then it is also as element of the algebra A^K algebraical over K by using the polynomial $t \, min_{a,K}$. If a is – as element of the algebra A – algebraical over K, then it is as element of A^K algebraical over K by using the same polynomial. Thus, we can use the word algebraical without no confusion in the context of A and A^K.$\diamond$

Corollary 14 Let A be unitary and $a \in A$ algebraical over K. The following statements are valid:

(i) If $a \in E(A)$ is valid, then $t \cdot min_{a,K} = \widetilde{min}_{a,K}$ is true. In this case t is no divisor of $min_{a,K}$.

(ii) If $a \in N(A)$ is valid, then $min_{a,K} = \widetilde{min}_{a,K}$ is true. In this case t is a divisor of $min_{a,K}$.

Proof. ad(i): $min_{a,K} \mid \widetilde{min}_{a,K} \mid t \cdot min_{a,K}$ is valid. Let $g, h \in K[t]$ such that $\widetilde{min}_{a,K} = min_{a,K} \cdot h$ and $t \cdot min_{a,K} = \widetilde{min}_{a,K} \cdot g$ are true. We deduce that g and h possess a leading coefficient 1 and that $t = hg$ is valid. If $h = 1_K$ would be true, then $\widetilde{min}_{a,K} = min_{a,K}$ would be valid. Based on corollary 13 we would deduce $t \mid min_{a,K}$. This is a contradiction to part (ii) of lemma 8 (with $A := K[a]$). Hence, part (i) is proven.

ad(ii): $min_{a,K} \mid \widetilde{min}_{a,K}$ is straightforward to prove. Based on corollary 13 (with $A := K[a]$) we conclude $\widetilde{min}_{a,K} \mid min_{a,K}.\diamond$

Within the next theorem we deduce that both concepts are compatible.

Theorem 44 Let A be unitary. The following statements are valid:

(i) $H(A) = H(A^K) \cap A$

(ii) $VSep(A) = VSep(A^K) \cap A$

(iii) $Sep(A) = Sep(A^K) \cap A$

Commutative algebras   183

(iv) $ZF(A) = ZF(A^K) \cap A$

(v) $D(A) = D(A^K) \cap A$

Proof. The theorem is a direct consequence of lemma 9 and corollary 14.⋄

Theorem 44 implies that the definitions within this section are compatible with the prior ones. Therefor, we use the same symbols within the next definition.

Definition and remark 11 Let $D(A)$, $ZF(A)$, $VSep(A)$, $Sep(A)$, $Nil(A)$ resp. $H(A)$ be the set of diagonizable, splittable, fully separable, separable, nilpotent resp. semisimple elements of A over K. By definition the following identities are valid:
$D(A) = D(A^K) \cap A$, $ZF(A) = ZF(A^K) \cap A$, $Sep(A) = Sep(A^K) \cap A$, $VSep(A) = VSep(A^K) \cap A$, $Nil(A^K) \cap A = Nil(A)$ and $H(A) = H(A^K) \cap A$. If $a \in A$, then we call a pair $(r; s) \in A \times A$ a generalized Jordan decomposition of a if the following conditions are valid: $a = r + s$, $rs = sr$, $r \in Nil(A)$ and $s \in VSep(A)$.⋄

We end this section by transferring all results for commutative unitary algebras to commutative non-unitary algebras. The first theorems transfers the results for $Sep(A)$ and $VSep(A)$, the second one the results for $ZF(A)$ and $D(A)$ and the last one the results for the generalized Jordan decomposition.

Theorem 45 *The following statements are valid:*

(i) *$a \in VSep(A)$ is valid if and only if $\langle a \rangle_A$ is separable.*

(ii) *$a \in Sep(A)$ is valid if and only if $\langle a \rangle_A$ is finite-dimensional and possesses a separable factor algebra by its nilradical.*

(iii) *If A is commutative, then $VSep(A)$ is a K-subalgebra of A.*

(iv) *Let A be commutative and finite-dimensional.*

  (a) *$Sep(A) = VSep(A) \oplus_K rad(A)$ is valid. In particular, $Sep(A)$ is a K-subalgebra of A.*

  (b) *$VSep(A)$ is a separable K-subalgebra of A and the unique algebra complement of $rad(A) = rad(Sep(A))$ in $Sep(A)$.*

  (c) *A is separable if and only if $A = VSep(A)$ is valid.*

  (d) *The following statements are equivalent:*

    (1) *$A/rad(A)$ is separable.*
    (2) *$Sep(A) = A$*
    (3) *$VSep(A)$ is an algebra complement of $rad(A)$ in A.*

(4) $VSep(A)$ is the unique algebra complement of $rad(A)$ in A.

(e) $VSep(A)$ is the unique maximal semisimple and separable K-subalgebra of $Sep(A)$.

(f) $VSep(A)$ is the unique maximal separable K-subalgebra of A.

(g) $Sep(A)$ is the unique maximal K-subalgebra of A possessing a separable factor algebra by its nilradical.

Proof. ad(i): Let $a \in VSep(A)$. By using definition 11 we deduce $a \in VSep(A^K)$. Hence, $K[a]$ is a separable subalgebra of A (see corollary 8). Based on part (ii) of remark 18 the subalgebra $\langle a \rangle_A$ is separable. If $\langle a \rangle_A$ is separable, then $K[a]$ is – by using definition and remark 10 and part (ii) of proposition 2 – separable, too. We use corollary 8 to deduce $a \in VSep(A^K) \cap A = VSep(A)$.

ad(ii): Let $a \in A$ be algebraical over K. Based on corollary 8 we deduce that $a \in VSep(A^K)$ is valid if and only if $K[a]/rad(K[a])$ is separable. Because of definition and remark 10 and part (iv) of corollary 3 this is equivalent to the fact that $\langle a \rangle_A / rad(\langle a \rangle_A)$ is separable.

ad(iii): Let A be commutative. Based on theorem 35 the set $VSep(A^K)$ is a subalgebra of A^K. Hence, $VSep(A) = VSep(A^K) \cap A$ is a subalgebra of A.

ad(iv): Let A be commutative and finite-dimensional.

(a): By using part (i) of theorem 36, part (ii) of corollary 3 and Dedekind's identity we deduce: $Sep(A) = Sep(A^K) \cap A = (VSep(A^K) \oplus_K rad(A)) \cap A = rad(A) \oplus_K (VSep(A^K) \cap A) = rad(A) \oplus_K VSep(A)$.

(c): Straightforward to prove is the statement $VSep(A^K) = VSep(A) \oplus_K K1_{A^K}$. In addition, by using $A^K = A \oplus_K K1_{A^K}$, part (iv) of theorem 36 and part (iv) of corollary 3 we deduce: $A = VSep(A) \iff A^K = VSep(A^K) \iff A^K$ separable $\iff A$ separable.

(b): This statement is deductable by using part (c) and part (vi) of theorem 16.

(d): Let (1) be valid. Based on part (iv) of corollary 3 we deduce that A^K possesses a separable factor algebra by its nilradical. Now we use part (v) of theorem 36 to deduce $A^K = Sep(A^K)$. $Sep(A^K) = Sep(A) \oplus_K K1_{A^K} \cong_{A_1} (K, Sep(A))$ is valid, and thus $Sep(A) = A$ is true. Let (2) be valid. By using the parts (iii) and (iv),(a) we deduce (3). The implication from (3) to (4) is straightforward to verify. Finally, from part (4), part (b) and part (vi) of theorem 16 we deduce statement (1).

(e) and (f): Because of part (b) the subalgebra $VSep(A)$ possesses the desired properties. Based on part (iv),(a) and (b) and part (ii) of theorem 26 every semisimple subalgebra of $Sep(A)$ is separable. Let T be a separable subalgebra of A. By using part (c) we obtain $T = VSep(T) \subseteq VSep(A)$.

(g): By using part (d) we deduce that $Sep(A)$ possesses the stated properties. Let T be a subalgebra of A possessing a separable factor algebra by its nilradical. We use part (d) to deduce $T = Sep(T) \subseteq Sep(A)$.⋄

Theorem 46 *Let $a \in A$. The following statements are valid:*

(i) $a \in D(A)$ is valid if and only if $\langle a \rangle_A$ is finite-dimensional, semisimple and K is a splitting field for $\langle a \rangle_A$.

(ii) $a \in ZF(A)$ is valid if and only if $\langle a \rangle_A$ is finite-dimensional and K is a splitting field for $\langle a \rangle_A$.

(iii) $a \in H(A)$ is valid if and only if $\langle a \rangle_A$ is finite-dimensional and semisimple.

(iv) If A is commutative, then $D(A)$ is a K-subalgebra of A. If $D(A)$ is commutative and finite-dimensional, then $D(A)$ is separable and K is a splitting field for $D(A)$.

(v) Let A be finite-dimensional and commutative. The following statements are valid:

   (a) $ZF(A) = D(A) \oplus_K rad(A)$ is valid. In particular, $ZF(A)$ is a K-subalgebra of A and $rad(ZF(A)) = rad(A)$ is true.

   (b) $D(A)$ is the unique algebra complement of $rad(A)$ in $ZF(A)$.

   (c) $D(A)$ is the unique maximal semisimple and separable K-subalgebra such that K is a splitting field for it. In particular, $A = D(A)$ if and only if A is $\mathcal{A}_1$-isomorph to $K^{dim_K(A)}$.

   (d) $ZF(A)$ is the unique maximal K-subalgebra of A such that K is a splitting field for it.

   (e) If $A/rad(A)$ is separable, then $H(A) = VSep(A)$ is valid.

Proof. ad(i) and (ii): Let $a \in ZF(A)$ resp. $a \in D(A)$. Based on the Chinese-Remainder theorem the subalgebra $K[a]$ is finite-dimensional and splittable over K resp. finite-dimensional, semisimple and splittable over K. We use part (iii) of theorem 26 to deduce that the subalgebra $\langle a \rangle_A$ of $K[a]$ possesses these properties, too. If, vice versa, K is a splitting field for the finite-dimensional (and resp. semisimple) algebra $\langle a \rangle_A$, then we use definition and remark 10 and part (ii) of corollary 3 that $K[a]$ is (resp. semisimple,) finite-dimensional and splitting over K. The Chinese-Remainder theorem lets us deduce that $a \in ZF(A^K) \cap A = ZF(A)$ resp. $a \in D(A^K) \cap A = D(A)$ is valid.

ad(iii): Let $a \in H(A)$. By using the Chinese-Remainder theorem we deduce that $K[a]$ is finite-dimensional and semisimple. Hence, the ideal $\langle a \rangle_A$ (see definition and remark 10) possesses this properties, too. Let, vice versa, $\langle a \rangle_A$ be finite-dimensional and semisimple. Based on definition and remark 10 and part (ii) of corollary 3 the same properties are valid for $K[a]$. Based on the Chinese Remainder theorem we deduce $a \in H(A^K) \cap A = H(A)$.

ad(iv): Based on theorem 39 the set $D(A^K)$ is a subalgebra of A^K. Hence, $D(A) = D(A^K) \cap A$ is a subalgebra of A. Straightforward to prove is $D(A)^K = D(A) \oplus_K K1_{A^K} \cong_{A_1} (K, D(A))$. Finally, we use corollary 3 to prove part (iv).

ad(v): (a): This part is a consequence of part (i) of theorem 40, part (ii) of corollary 3 and Dedekind's identity.

(b): This part is deductable from part (iv), (v),(a) and part (vi) of theorem 16.

(c): Based on part (iv) and part (i) of corollary 1 the subalgebra $D(A)$ possesses the desired properties. Let T be a semisimple subalgebra of A splitting over K. We use part (ii) of proposition 2 to deduce that T is separable. If $t \in T$, then part (iii) of theorem 26 is used to deduce that $\langle t \rangle_A$ is finite-dimensional, semisimple and splitting over K. By using this statement and part (i) we conclude $t \in D(A)$.

(d) Based on part (v),(a) and part (iv) the subalgebra $ZF(A)$ possesses the desired properties. Let T be a subalgebra of A such that K is a splitting field for it and $t \in T$. We use part (iii) of theorem 26 to deduce that $\langle t \rangle_A$ splits over K. Thus, by using this statement and part (ii) we conclude $t \in ZF(A)$.

(e) Let $A/rad(A)$ separable. Based on part (iv) of corollary 3 the algebra $A^K/rad(A^K)$ is separable. We use part (iv) of theorem 40 to deduce $H(A^K) = VSep(A^K)$. An intersection with A proves the theorem.◇

Theorem 47 *Let $a \in A$. The following statements are valid:*

(i) The following statements are equivalent:

   (a) $a \in Sep(A)$

Commutative algebras   187

   (b) a possesses a generalized Jordan decomposition in $\langle a \rangle_A$.

   (c) a possesses a generalized Jordan decomposition in A.

(ii) a possesses at most one generalized Jordan decomposition in A.

(iii) Let $(r; s)$ be a generalized Jordan decomposition of a in $\langle a \rangle_A$ and $f, g \in tK[t]$ such that $r = f(a)$ and $s = g(a)$.

   (a) $cl(r) = max(\widetilde{min}_{a,K})$

   (b) $\widetilde{min}_{s,K} = halb(\widetilde{min}_{a,K})$

   (c) $\widetilde{min}_{s,K} \mid \widetilde{min}_{a,K} \mid \widetilde{min}_{s,K}^{cl(r)}$

   (d) Let $p, q \in tK[t]$. $(p(a); q(a))$ is a generalized Jordan decomposition of a in A if and only if $p \equiv f \mod \widetilde{min}_{r,K}$ and $q \equiv g \mod \widetilde{min}_{s,K}$ are valid.

   (e) Let ρ be the right regular representation of $K[a]$. The summands of the primary decomposition of A^K into $K[a\rho]$-modules are – with respect to $K[s\rho]$-isomorphism – exactly the irreducible $K[s\rho]$-modules of the $K[s\rho]$-module A^K. For every summand W of the primary decomposition of A^K into $K[a\rho]$-modules a basis B_W of W exists such that $M_B(r\rho_{|W})$ is a strict lower triangular matrix.

   (f) $a \in ZF(A)$ is valid if and only if $s \in D(A)$ is true.

Proof. ad(i): Let (a) be valid. Based on part (i) of theorem 38 the element a possesses a generalized Jordan decomposition in $K[a]$. By using definition 10 and part (ii) of corollary 3 we deduce the statement $rad(K[a]) = rad(\langle a \rangle_A)$. Thus, part (b) is valid. The implication from (b) to (c) is straightforward to prove. Finally, let (c) be valid. Based on part (i) of theorem 38 we deduce $a \in Sep(A^K) \cap A = Sep(A)$.

ad(ii): Every generalized Jordan decomposition of a in A is also one of a in A^K. Thus, part (ii) is deductable from part (ii) of theorem 38.

ad(iii): (a),(b): The argumentation is similar as done in part (ii). In this case the parts (a) and (b) are deductable from part (iii),(a) and (b) of theorem 38 and from corollary 7.

(c): This part is a consequence of the parts (a) and (b).

(d): The argumentation is similar as done in part (ii).

(e): This statement is a consequence of part (v),(a) of theorem 46.◇

5.8 Solvable algebras

Within this final section we answer the question on what terms $VSep(A)$ is a radical complement. Those algebras are related to Lie nilpotent algebras. For analyzing this topic the next proposition is relevant. The answer is used for describing the radical complements of solvable associative algebras by using fully separable elements. This description is used within the context of Cartan subalgebras of associated Lie algebras which is analyzed in the work [53].

Proposition 15 *(radical and nilpotent elements) Let A be a right artian associative K-algebra. The following statements are equivalent:*

(i) $rad(A) = Nil(A)$

(ii) $Nil(A/rad(A)) = \{0_A + rad(A)\}$

(iii) $A/rad(A)$ is $\mathcal{A}_1$-isomorphic to a direct sum of K-division algebras.

Proof. Let $rad(A) = Nil(A)$ be valid. If $a \in A$ and $a+rad(A)$ is nilpotent, then an element $n \in \mathbb{N}$ exists such that $a^n \in rad(A)$ is valid. $rad(A)$ is nil, and thus an element $m \in \mathbb{N}$ exists such that $(a^n)^m = 0_A$ is true. We conclude that a is nilpotent, and based on $rad(A) = Nil(A)$ we deduce $a \in rad(A)$. Let $Nil(A/rad(A)) = \{0_A + rad(A)\}$ be valid. If $a \in Nil(A)$, then $a + rad(A) \in Nil(A/rad(A))$ is valid. By using the assumption we conclude $a \in rad(A)$. Hence, the parts (i) and (ii) are equivalent.

By using the theorem of Wedderburn-Artin the algebra $A/rad(A)$ is $\mathcal{A}$-isomorphic to a direct sum of complete matrix algebras over K-division algebras. If $n \in \mathbb{N}_{\geq 2}$ and D is a K-division algebra, then the algebra $D^{n \times n}$ possesses nilpotent elements different from zero. A direct sum of K-division algebras possesses no nilpotent elements different from zero. We conclude that the parts (ii) and (iii) are equivalent.◇

Theorem 48 *Let K be a perfect field and A a finite-dimensional associative K-algebra. The following statements are equivalent:*

(i) $VSep(A)$ is a radical complement.

(ii) Exactly one radical complement exists, and $A/rad(A)$ is $\mathcal{A}_1$-isomorphic to a direct sum of K-division algebras.

Proof. Let $VSep(A)$ be a radical complement. K is perfect, and thus based on theorem 15 and remark 9 the set of all radical complements is exactly the orbit of $VSep(A)$ under the action of $(rad(A); *)$ by conjugation. $VSep(A)$ is invariant under this action. we conclude that $VSep(A)$ is the unique radical complement. $Nil(VSep(A)) = \{0_A\}$ is valid, and thus

$Nil(A/rad(A)) = \{rad(A)\}$ is true. Proposition 15 is used to prove part (ii).
Now let part (ii) be valid. Let T be a radical complement. If $x \in VSep(A)$, then we use the perfectness of K and part (i) of theorem 45 to deduce that $\langle x \rangle_A$ is a separable subalgebra of A. Again, by using the perfectness of K this subalgebra is contained based on part (i) of corollary 4 in a radical complement of A. We have proven $VSep(A) \subseteq T$. Let $t \in T$. K is perfect, and thus $t \in Sep(A)$ is valid. We use part (i) of theorem 47 and elements $r \in Nil(A)$ and $x \in VSep(A)$ such that $t = r + x$ is valid. By the previous implication already proven we deduce $r \in Nil(T)$. Proposition 15 is used to deduce $Nil(A/rad(A)) = \{rad(A)\}$. Hence, $Nil(T) = \{0_A\}$ is valid, and we conclude $r = 0_A$. Hence, $t = x \in VSep(A)$ is valid, and we have proven $T = VSep(A)$.⋄

Corollary 8 *Let K be a perfect field and A a finite-dimensional associative solvable K-algebra. The following statements are equivalent:*

(i) $VSep(A)$ is a radical complement.

(ii) Exactly one radical complement exists.

Proof. A is solvable, and thus $A/rad(A)$ is $\mathcal{A}_1$-isomorphic to a direct sum of fields. The statements of theorem 48 are used to finish the proof.⋄

Now we answer the question in what way radical complements for solvable associative algebras are related to the set of fully separable elements.

Proposition 16 *Let K be a field, A a finite-dimensional associative solvable K-algebra, $A/rad(A)$ separable and T a K-subalgebra of A. $T \subseteq VSep(A)$ is valid if and only if T is commutative and separable.*

Proof. T is commutative and separable, and thus $T = VSep(T)$ is valid based on part (iv),(c) of theorem 46. In particular, $T \subseteq VSep(A)$ is true. Let $T \subseteq VSep(A)$ be valid. Because of $Nil(A) \cap VSep(A) = \{0_A\}$ the algebra T possesses no nilpotent elements different from 0_A. Thus, T is semisimple, and based on part (ii) of theorem 26 we deduce the separability of T. In particular, we use part (i) of corollary 4 to prove that T in a radical complement of A. A is solvable, and thus this complement is commutative. Thus, T is commutative, too.⋄

The following theorem implies that for solvable associative algebras the radical complements are maximal separable subalgebras. Within the context of associated Lie algebras analyzed in the work [53] this theorem is reformulated: radical complements are so-called maximal tori.

Theorem 49 Let K be a field, A a finite-dimensional associative solvable K-algebra, $A/rad(A)$ separable and T a subset of A. The following statements are equivalent:

(i) T is a radical complement of A.

(ii) T is – with respect to all K-subalgebras contained in $VSep(A)$ – a $\subseteq$-maximal element.

Proof. Let T be a radical complement of A. T is commutative and separable. Proposition 16 is used to deduce $T \subseteq VSep(A)$. Let H be a subalgebra of A such that $T \subseteq H \subseteq VSep(A)$ is valid. H is – by using proposition 16 – a separable subalgebra of A. We use part (i) of corollary 4 to find an element $r \in rad(A)$ such that $H^{(r)} \subseteq T$ is valid. By using $dim_K(H) = dim_K(H^{(r)})$ (see remark 9) we deduce $T = H$.

Let T be a $\subseteq$-maximal element of $VSep(A)$ with respect to all subalgebras of A. $A/rad(A)$ is separable, and we use theorem 13 to find a radical complement X in A. Proposition 16 is used to deduce that T is a separable subalgebra of A. Based on part (i) of corollary 4 an element $r \in rad(A)$ exists such that $T^{(r)} \subseteq X$ is true. We use remark 9 to prove that T and $T^{(r)}$ are $\subseteq$-maximal elements of $VSep(A)$ with respect to all subalgebras of A. Hence, $T^{(r)} = X$ is valid, and based on remark 9 the set T is a radical complement in A. ⋄

Based on this theorem we are able to analyze two extreme cases for the set $VSep(A)$ within solvable associative algebras.

Corollary 9 Let K be a field, A a finite-dimensional associative solvable K-algebra and $A/rad(A)$ separable. The following statements are valid:

(i) $A = VSep(A)$ is valid if and only if A is commutative and separable.

(ii) $VSep(A)$ is a radical complement if and only if $VSep(A)$ is a K-subalgebra of A.

(iii) Let K be perfect. The following statements are equivalent:

   (a) Exactly one radical complement exists.

   (b) $VSep(A)$ is the unique radical complement.

   (c) $VSep(A)$ is a radical complement.

   (d) $VSep(A)$ is a K-subalgebra of A.

   (e) $VSep(A)$ is a K-subspace of A.

   (f) $VSep(A)$ is central. In particular, A is Lie nilpotent.

Proof. ad(i): This part is valid by using theorem 49 and proposition 16.

ad(ii): The equivalence is valid based on theorem 49.

ad(iii): The equivalence of the parts (a), (b) and (c) is proven within corollary 8. In addition, the parts (b) and (d) are equivalent based on part (ii). Straightforward to prove is that statement (d) implies (c). We prove now the implication from part (e) to (b). Let $VSep(A)$ be a K-subspace of A. K is perfect, and thus based on theorem 13 a radical complement T exists. This complement is based on theorem 49 contained in $VSep(A)$. $VSep(A) \cap rad(A) = \{0_A\}$ is valid, and thus comparing dimension we conclude $T = VSep(A)$. If part (b) is valid, then we use corollary 6 to prove that $VSep(A)$ is central (because the intersection of all radical complements is central). Let $VSep(A)$ be central. We deduce $VSep(A) = VSep(Z(A))$, and based on theorem 35 we conclude that $VSep(Z(A))$ is a subalgebra of $Z(A)$. Hence, $VSep(A)$ is a subalgebra of $Z(A)$. If $VSep(A)$ is a central radical complement, then the Lie powers of A are exactly the Lie powers of $rad(A)$. The latter ideal is associative nilpotent, and hence it is also Lie nilpotent. Therefor, A is Lie nilpotent.⋄

We focus on the lower triangular matrix to obtain an example linked to this theorem.

Example 17 Let $A := \Delta_{u,2}$ and $e, r \in A$ such that $rad(A) = \langle r \rangle_K$ is valid and $C := \langle 1_A, e \rangle_K$ is a radical complement. We calculate $r^2 = 0_A$, and thus $VSep(A) \neq A$ is true. Within theorem 16 we have already proven that C possesses maximal orbit length. In particular, at least two radical complements exist. If K is perfect, then we use corollary 9 to deduce that $VSep(A)$ is no K-subspace of A. Indeed, a straightforward calculation leads to: $VSep(A) = K1_A \cup \{a \mid \exists k, m \in K, l \in K \setminus \{0_K\} : a = k1_A + le + mr\}$.⋄

The last theorem within this work presents an idea for determining a radical complement for solvable associative algebras.

Theorem 50 *Let K be a field, A a finite-dimensional associative solvable K-algebra and $A/rad(A)$ separable. The following statements are valid:*

(i) If B is – with respect to all linear independent subsets of $VSep(A)$ such that their elements commute pairwise – a $\subseteq$-maximal element, then $\langle B \rangle_K$ is a radical complement.

(ii) If B is a basis of a radical complement of A, then B is – with respect to all linear independent subsets of $VSep(A)$ such that their elements commute pairwise – a $\subseteq$-maximal element

Proof. ad(i): Let $C := \langle B \rangle_A$. C is a finite-dimensional commutative K-algebra. By using our assumption $B \subseteq VSep(C)$ is valid, and we use part (iv),(b) of theorem 45 to prove $C = VSep(C)$. In particular, C is separable and $C \subseteq VSep(A)$ is true. Let X be a subalgebra of A such that $C \subseteq X \subseteq VSep(A)$ is valid. Based on proposition 16 the set X is a commutative separable K-algebra. The linear algebraic concept of enhancing a linear independent set to a basis and the maximality of B are used to prove $X = C$. Thus, C is a radical complement based on theorem 49. C is commutative and contained in $VSep(A)$, and therefor (again using the linear algebraic concept of enhancing a linear independent set to a basis and the maximality of B) the statement $\langle B \rangle_K = C$. Thus, part (i) is proven.

ad(ii): Let B be a basis of a radical complement C. By using theorem 49 the set B is a linear independent subset of $VSep(A)$ possessing pairwise commuting elements. Let T be a linear independent subset of $VSep(A)$ such that its elements commute pairwise and that $B \subseteq T$ is valid. Straightforward to prove is the statement $\langle B \rangle_A \subseteq \langle T \rangle_A$. We use part (iv),(b) of theorem 45 to deduce that $\langle T \rangle_A$ is a separable subalgebra of A. Thus, by using part (iv) of corollary 4, the statement $\langle B \rangle_A = \langle T \rangle_A$ is true. Now we use part (i) to deduce $\langle T \rangle_K \subseteq \langle T \rangle_A = \langle B \rangle_A = \langle B \rangle_K$. Hence, $\mid T \mid \leq \mid B \mid$ and $B = T$ are valid.⋄

Pure mathematics is, in its way, the poetry of logical ideas. (Albert Einstein)

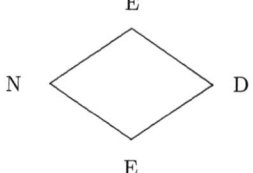

5.9 Open-ended questions and exercises

Open-ended question 3 *(i) What is the importance of the values $dim_K(A) - dim_K(Sep(A))$ and $dim_K(A) - dim_K(VSep(A))$?*

(ii) Let K be a field and M a commutative monoid or magma. Determine the structure of KM.

(iii) Let $a = r + t$ be a generalized Jordan decomposition. Is it possible to determine $min_{a,K}$ based on $min_{r,K}$ and $min_{t,K}$?

5.10 Exercises

Excercise 282 *Every commutative associative algebra is the center of an associative commutative algebra.*

Excercise 283 *Let A be an associative algebra and I a nilpotent ideal of A. Prove that a is fully-separable if and only if $a + I$ is fully-separable and both possess the same minimum polynomial.*

Excercise 284 *The calculations $\frac{2}{8} = \frac{1}{4}$ and $\frac{2+1}{8+1} = \frac{3}{9} = \frac{1}{3} = \frac{1}{4-1}$ are valid in $\mathbb{Q}$. Within $\mathbb{Q}$ prove or disprove that for all $x, n \in \mathbb{N}$ elements $a, b \in \mathbb{N}$ exists such that $\frac{a}{b} = \frac{1}{x}$ and $\frac{a+n}{b+n} = \frac{1}{x-n}$ is valid. Find all solutions for this equations. Is it possible to generalize this statements to quotient fields of rings or to general associative algebras?*

Excercise 285 *Let K be a field and M a commutative monoid such that $e^2 = e$ for all $e \in M$ is valid. Determine the structure of KM. Find examples for M.*

Excercise 286 *Prove that the tensor product of two separable algebras is separable by using the characterization of separating idempotents. Do the same for the direct sum of two separable algebras.*

Excercise 287 *Prove that the tensor product $A \otimes B$ of a separable algebra A with a semisimple algebra B is semisimple. (Hint: First reduce the exercise to the case that A is simple and separable and B is simple. Let L be the center of B. L is a field. Use $A \otimes_K B \cong A \otimes_K (L \otimes_L B) \cong (A \otimes_K L) \otimes_L B$. A is separable, and thus $(A \otimes_K L)$ is semisimple. B is central-simple over L. Now apply a theorem within the theory of central-simple algebras: the tensor product of simple and central-simple algebra is simple.)*

Excercise 288 *This exercise is taken from [8], chapter 6. Prove the following theorem of Dickson for finite-dimensional associative central division algebras D over fields K: Two elements $a, b \in D$ are conjugate if and only if they possess the same minimal polynomial: $min_{a,K} = min_{b,K}$. (Hint: Apply the theorem of Skolem-Noether to the algebra generated by a and b. Afterwards use a well-known theorem of the theory of fields about algebraic conjugate elements and homomorphism of the Galois group.) Use the theorem of Dickson and present some conjugate and non-conjugate elements of the real quaternion algebra.*

Excercise 289 Let K be a field. We focus on a cyclic group C_n of order n. Prove that the group algebra is isomorphic to $K[t]/(t^n - 1)$. If $t^n - 1 = \prod_k f_k(t)^{m_k}$ is the decomposition of $t^n - 1$ over $K[t]$ into irreducible polynomials, then apply the Chinese-Remainder theorem to decompose the group algebra into fields. Is there a special result for $K = \mathbb{Q}$? (Hint: see [60])

Excercise 290 Let K be a field, A a finite-dimensional associative solvable K-algebra and $A/rad(A)$ separable. True or false: $VSep(A)$ is the sum of all radical complements of A.

Excercise 291 Read the following text written by Geoffrey Robinson (see [62]) and do a research in the literature for understanding its content. Within this text the irreducible modules of semisimple commutative group algebras are constructed:

> This is all fairly standard, but here goes. If M is an irreducible KG-module, then (since G is Abelian) we obtain a homomorphism $\theta : G \to \mathrm{End}_{KG}(M)^\times$, so $\theta(G)$ is a finite Abelian subgroup of the group of units of a division algebra (using Schur's Lemma). Hence $\mathrm{Im}\theta$ is cyclic. In other words, the problem is now reduced to proving the result in the case G cyclic. Let us recalibrate notation, and assume that $|G| = r$ and $G = \langle g \rangle$ is cyclic. Let us consider out irreducible KG module M and θ as before. Let $\theta(g)$ have order d (we could actually assume that $d = r$ at this point, given what has gone before, but let us work in greater generality). Let $f(x) \in K[x]$ be the minimum polynomial of $\theta(g)$. Then $f(x)$ must be irreducible, for if $p(x) \in K[x]$ is an irreducible factor of $f(x)$, then $p(\theta(g))$ is not invertible, so must be the zero matrix, as it commutes with all of $\theta(G)$. Also, $f(x)$ divides $x^d - 1$, since $\theta(g)$ has order d, and $f(x)$ does not divide $x^h - 1$ (hence is coprime to it) for $0 < h < d$. Each eigenvalue of $\theta(g)$ (in a suitable extension of K) is therefore a primitive d-th root of unity. Let ω_d be one of these. By the theory of the rational canonical form, we may choose a K-basis for M such that the matrix X representing $\theta(g)$ with respect to that basis is the companion matrix for $f(x)$. The size of the matrix for $\theta(g)$ is thus $[K(\omega_d) : K] \times [K(\omega_d) : K]$. For each positive integer a coprime to d, and less than d, we may define a new matrix representation ψ of $\langle g \rangle$ by setting $\psi(g) = X^a$ (and following the previous constructions carefully, (up to similarity) there are no other choices for which g can act as a matrix of order exactly d on an irreducible KG-module). This explains how to construct all $\phi(d)$ inequivalent irreducible representations of $\langle g \rangle$ of degree $[K(\omega_d) : K]$ over K (where ϕ is Euler's function).

Commutative algebras 195

Excercise 292 *Let K be a field and $n \in \mathbb{N}$. True or false: the center of the following algebras is separable:*

(i) $\mathbb{H}$ *as $\mathbb{R}$-algebra*

(ii) K *as K-algebra*

(iii) $\mathbb{R}$ *as $\mathbb{Q}$-algebra*

(iv) $K^{n \times n}$ *as K-algebra*

(v) $A^{n \times n}$ *as K-algebra for a separable K-algebra A*

(vi) $A^{n \times n}$ *as K-algebra for a separable commutative K-algebra A*

(vii) $A(a, b, K)$ *for $char(K) \neq 2$*

(viii) $A(a, b, K)$ *for $char(K) = 2$*

(ix) $\Delta_{u,n}$ *as $\mathbb{R}$-algebra*

(x) $\Delta_{o,n}$ *as $\mathbb{Q}$-algebra*

(xi) $\mathbb{H} \times \mathbb{R} \times \mathbb{R}^{2 \times 2}$ *as $\mathbb{R}$-algebra*

(xii) *a complex finite-dimensional associative unitary algebra*

(xiii) *a complex finite-dimensional associative commutative unitary algebra*

(xiv) *a complex finite-dimensional associative unitary semisimple algebra*

(xv) *a complex finite-dimensional associative unitary commutative semisimple algebra*

(xvi) *a nilpotent complex algebra*

(xvii) *a nilpotent algebra*

(xviii) $A \times B$ *for associative finite-dimensional unitary K-algebras A, B possessing a separable factor algebras by their nilradicals*

(xix) $A \times B$ *for associative finite-dimensional unitary solvable K-algebras A, B possessing a separable factor algebras by their nilradicals*

(xx) $A \otimes B$ *for associative finite-dimensional unitary K-algebras A, B possessing a separable factor algebras by their nilradicals*

(xxi) $A \otimes B$ *for associative finite-dimensional unitary solvable K-algebras A, B possessing a separable factor algebras by their nilradicals*

Excercise 293 *Within exercise 292 determine – if possible – all elements of the center explicitly (intrinsic description).*

Excercise 294 *Within exercise 292 determine – if possible – all elements of the center by using the covering algebra (external description).*

Excercise 295 *Within exercise 292 determine – if possible – for all central elements the generalized Jordan-decomposition.*

Excercise 296 *Within exercise 292 determine – if possible – the nilradical and the unique radical complement of the center of the algebra.*

Excercise 297 *Within exercise 292 determine – if possible – the subalgebras of the separable, fully separable, semisimple, nilpotent, splittable and diagonalizable elements.*

Excercise 298 *Within exercise 292 draw – if possible – a Hasse diagram for the subalgebras of all separable, fully separable, semisimple, nilpotent, splittable and diagonalizable elements.*

Excercise 299 *Within exercise 173 analyze the center of eAe with respect to the following topics:*

   (i) Determine the elements of the center by an intrinsic description.

   (ii) Determine the elements of the center by an external description.

   (iii) Determine the nilradical and one radical complement.

   (iv) Does only one radical complement exist?

   (v) Determine the Jordan-decomposition of an arbitrary element.

   (vi) On what terms is a central element semisimple?

   (vii) On what terms is a central element nilpotent?

  (viii) On what terms is a central element separable?

   (ix) On what terms is a central element fully separable?

    (x) On what terms is a central element diagonalizable?

   (xi) On what terms is a central element splittable?

   (xii) Draw Hasse diagrams for the subalgebras of all separable, fully separable, semisimple, nilpotent, splittable and diagonalizable elements of the algebra and the center of the algebra.

If the exercise is too complex, then solve the exercise under the precondition that e is central.

Commutative algebras

Excercise 300 *Within 174 analyze the center of the zero-extension with respect to the following topics:*

(i) *Determine the elements of the center by an intrinsic description.*

(ii) *Determine the elements of the center by an external description.*

(iii) *Determine the nilradical and one radical complement.*

(iv) *Does only one radical complement exist?*

(v) *Determine the Jordan-decomposition of an arbitrary element.*

(vi) *On what terms is a central element semisimple?*

(vii) *On what terms is a central element nilpotent?*

(viii) *On what terms is a central element separable?*

(ix) *On what terms is a central element fully separable?*

(x) *On what terms is a central element diagonalizable?*

(xi) *On what terms is a central element splittable?*

(xii) *Draw Hasse diagrams for the subalgebras of all separable, fully separable, semisimple, nilpotent, splittable and diagonalizable elements of the algebra and the center of the algebra.*

If the exercise is too complex, then solve the exercise under the precondition that A is separable or separable and commutative.

Excercise 301 *Let K be a field and A a finite-dimensional associative solvable K-algebra possessing a separable factor algebra by its nilradical. If the center of A is semisimple, then it is separable and, in addition, it is identical to the intersection of all radical complements of A. Find examples for this statement.*

Excercise 302 *On what terms is exercise 301 applicable to exercise 173?*

Excercise 303 *On what terms is exercise 301 applicable to exercise 174?*

Excercise 304 *Determine the nilradical of the algebra within example 12!*

Excercise 305 *Every idempotent element of an associative algebra is fully separable.*

Excercise 306 *Let A be an associative finite-dimensional K-algebra possessing a basis consisting of idempotent elements. A is the sum of all radical complements.*

Excercise 307 Let A be an associative finite-dimensional commutative K-algebra possessing a basis based on idempotent elements. A is diagonalizable.

Excercise 308 Let A be an associative finite-dimensional K-algebra possessing a basis based on fully separable elements. A is the sum of all radical complements.

Excercise 309 Let A be an associative finite-dimensional K-algebra possessing a separable factor algebra by its nilradical. True or false: If A is the sum of all radical complements, then A is possessing a basis consistent of fully separable elements. Find examples for this statement.

Excercise 310 Let A be an associative finite-dimensional K-algebra possessing a separable factor algebra by its nilradical. What structural properties does A possess, if the sum of all radical complements is a radical complement. Find examples for this statement.

Excercise 311 Is it possible to apply exercise 309 to exercise 173?

Excercise 312 Is it possible to apply exercise 309 to exercise 174?

Excercise 313 Let A be an associative finite-dimensional K-algebra possessing a separable factor algebra by the nilradical. Analyze for the sum of all radical complements on what terms it is a K-subalgebra, a K-subspace, a K-ideal, the whole algebra, exactly one radical complement, a K-left ideal or a K right ideal.

Excercise 314 For a field K and $a, b \in K$ we focus on the K-algebra $A := A(a, b, K)$. Analyze for the mentioned elements a, b how many pairwise non-isomorphic algebras arise:

(i) $K = GF(2)$, $a, b \in K$ arbitrary

(ii) $K = GF(4)$, $a, b \in K$ arbitrary

(iii) $K := GF(2)(t)$, $a, b \in \{0, 1, t, t^2, t + t^2, t^3\}$

(iv) $K := GF(2)(t_1, t_2)$, $a, b \in \{0, 1, t_1, t_2, t_1 + t_2, t_1{}^2, t_2{}^2, t_1{}^3, t_1 t_2\}$.

Excercise 315 Within exercise 314 analyze the following topics:

(i) Determine a decomposition of the algebra into the nilradical and one radical complement.

(ii) Is the determined radical complement in (i) unique?

(iii) Determine the generalized Jordan decomposition for an arbitrary element of the algebra.

Commutative algebras

Excercise 316 *Within exercise 314 analyze the following topics:*

(i) *Determine $H(A)$ and its type of isomorphism.*

(ii) *Determine $D(A)$ and its type of isomorphism.*

(iii) *Determine $VSep(A)$ and its type of isomorphism.*

(iv) *Determine $rad(A)$ and its class of nilpotency.*

(v) *Determine $Sep(A)$, its nilradical and one radical complement. Does another radical complement exist?*

(vi) *Determine $ZF(A)$, its nilradical and one radical complement. Does another radical complement exist?*

Excercise 317 *Let K be a field, $n \in \mathbb{N}$, A, B finite-dimensional associative unitary K-algebras, e a central idempotent of A, M a finite monoid and G a finite group. For the following algebras C analyze whether the statement $Nil(C) = rad(C)$ is valid:*

(i) $\Delta_{u,n}$

(ii) $\Delta_{o,n}$

(iii) KM for M being commutative and idempotent

(iv) KG for G being Abelian

(v) KG for $G = Q_8$ and $K := \mathbb{Q}$

(vi) KG for $G = D_8$ and $K := \mathbb{R}$

(vii) KG for $G = SD_8$ and $K := \mathbb{C}$

(viii) $K^{n \times n}$

(ix) $\mathbb{H}$

(x) eAe such that $Nil(A) = rad(A)$ is valid (see exercise 173)

(xi) zero-extension of A such that A is separable and $Nil(A) = rad(A)$ is valid (see exercise 174)

(xii) $A(a, b, K)$ for $char(K) = 2$

(xiii) $A(a, b, K)$ for $char(K) \neq 2$

If $Nil(C) = rad(C)$ is not valid, then analyze on what terms $Nil(C)$ is a K-subspace. What is the relevance of this fact for the K-algebra C?

Excercise 318 Let K be a field, $a, b \in K$, $n \in \mathbb{N}$, A, B finite-dimensional associative unitary K-algebras, e a central idempotent of A, M a finite monoid and G a finite group. For the following algebras C analyze whether $VSep(C)$ is a K-subspace:

(i) $\Delta_{u,n}$

(ii) $\Delta_{o,n}$

(iii) KM for M being commutative and idempotent

(iv) KG for G being Abelian

(v) KG for $G = Q_8$ and $K := \mathbb{Q}$

(vi) KG for $G = D_8$ and $K := \mathbb{R}$

(vii) KG for $G = SD_8$ and $K := \mathbb{C}$

(viii) $K^{n \times n}$

(ix) $\mathbb{H}$

(x) eAe such that $VSep(A)$ is a K-subspace (see exercise 173)

(xi) zero-extension of A such that A is separable and $VSep(A)$ is a K-subspace (see exercise 174)

(xii) $A(a, b, K)$ for $char(K) = 2$

(xiii) $A(a, b, K)$ for $char(K) \neq 2$

If $VSep(C)$ is no K-subspace, then analyze on what terms $VSep(C)$ is a K-subspace. What is the relevance of this fact for the K-algebra C?

Excercise 319 Within exercise 314 determine the units and divisors of zero for all mentioned algebras.

Excercise 320 Let $K := GF(2)$, $G = \mathbb{Z}_3$ and $H = \mathbb{Z}_4$. Determine the units and divisors of zero for the following algebras:

(i) KG

(ii) KH

(iii) $KG \otimes KG$

(iv) $KH \times KH$

(v) $K(G \times H)$.

Commutative algebras

Excercise 321 Let $K := \mathbb{Q}$ and $a \in K$. For the matrix $M := \begin{pmatrix} a & 0_K & 0_K \\ a & 1_K & 0_K \\ a & 1_K & 1_K \end{pmatrix}$
analyze the following topics:

(i) Determine the minimal and characteristical polynomial of M.

(ii) On what terms is M separable?

(iii) If M is separable, then calculate the generalized Jordan decomposition of M. In this case determine the minimal polynomial of the nilpotent and fully separable part.

(iv) On what terms is M splittable?

(v) If M is splittable, then calculate the generalized Jordan decomposition of M. In this case determine the minimal polynomial of the nilpotent and fully separable part.

(vi) On what terms is M nilpotent?

(vii) On what terms is M fully separable?

(viii) On what terms is M diagonalizable?

What are the results after changing the base field to $\mathbb{R}$, $\mathbb{C}$ or to a finite field?

Excercise 322 Generalize part (ii) of example 13 by using an arbitrary element $a \in K$ and the matrix $M := \begin{pmatrix} a & 0_K & 0_K \\ 1_K & 1_K & 0_K \\ 1_K & 1_K & a \end{pmatrix}$.

Excercise 323 For the following matrix $M := \begin{pmatrix} 1 & 1 & 0 & 0 \\ 0 & 2 & 0 & 0 \\ 0 & 0 & 2 & 1 \\ 0 & 0 & 2 & 3 \end{pmatrix}$ do the same calculation as described in example 14. What is the relevance of the field K for this calculations?

Excercise 324 Within exercise 323 and example 14 the determination of the minimal polynomial is done based on invariant subspaces. For this prove the following statements (see e.g. in [37], pages 280ff.) for a linear endomorphism f of a K-space V and f-invariant subspaces $U, U_1, \cdots, U_n$:

(i) $min_{f_{|U}, K} \mid min_{f, K}$

(ii) $min_{f_{V/U}, K} \mid min_{f, K}$

(iii) $min_{f, K} \mid min_{f_{|U}, K} \cdot min_{f_{V/U}, K}$

(iv) If V is the direct sum of the subspaces U_i, then
$min_{f,K} = kgV(min_{f_{|U_1},K}, \cdots, min_{f_{|U_n},K})$ is valid.

(v) $char_{f,K} = char_{f_{|U},K} \cdot char_{f_{V/U},K}$

(vi) If V is the direct sum of the subspaces U_i, then
$char_{f,K} = char_{f_{|U_1},K} \cdots char_{f_{|U_n},K}$ is valid.

Describe the relevance of these statements for the mentioned exercise and example.

Excercise 325 Let f be a linear endomorphism of a finite-dimensional K-space V of K-dimension n. Prove that $min_{f,K} \mid char_{f,K} \mid min_{f,K}^n$. Deduce that every irreducible factor of $char_{f,K}$ is also an irreducible factor of $min_{f,K}$ and vice versa. Apply this result to an associative algebra and its algebraical elements using the right and left regular representation.

Excercise 326 Let K be a field, $char(K) = 2$, $a, b \in K$ and $A := A(a, b, K)$. For the elements

(i) 1

(ii) i

(iii) j

(iv) k

(v) ai

(vi) aj

(vii) ak

(viii) $i + bk$

(ix) $ai + bj$

(x) $1 + i$

(xi) $1 + i + j + k$

determine whether it is

(a) semisimple

(b) fully separable

(c) separable

(d) nilpotent

Commutative algebras

(e) diagonalizable and/or

(f) splittable.

In addition, determine the generalized Jordan decomposition of these elements. How do the results change by assuming $char(K) \neq 2$?

Excercise 327 In this exercise we focus for a polynomial f over a field K on the factor algebra $A_{f,K} := K[t]/(f)$. For the following combinations of f and K

(i) $K := GF(2)$; $f := (t+1)(t-1)$

(ii) $K := GF(2)$; $f := t^2$

(iii) $K := GF(2)$; $f := (t+1)$

(iv) $K := GF(2)$; $f := t^2 + t + 1$

(v) $K := GF(3)$; $f := (t+1)(t-1)$

(vi) $K := GF(3)$; $f := t^2$

(vii) $K := GF(3)$; $f := (t+1)$

(viii) $K := GF(3)$; $f := (t+a)(t-1)$ for an element $a \in K$ such that $a \neq 0$ and $a \neq 1$ are true

(ix) $K := GF(3)$; $f = t^2 + t + a$ for an element $a \in K$ such that $a \neq 0$ and $a \neq 1$ are valid

(x) $K := \mathbb{Q}$; $f := (t+1)(t-1)$

(xi) $K := \mathbb{Q}$; $f := t^2$

(xii) $K := \mathbb{Q}$; $f := (t+1)$

(xiii) $K := \mathbb{Q}$; $f := t^2 + t + 1$ and

(xiv) $K := \mathbb{R}$; $f := t^2 + \sqrt{(2)}t + 1$

determine for $A_{f,K}$ the subalgebras

(a) $H(A_{f,K})$

(b) $D(A_{f,K})$

(c) $Sep(A_{f,K})$

(d) $VSep(A_{f,K})$

(e) $rad(A_{f,K})$ and

(f) $ZF(A_{f,K})$.

Visualize the results for each example by using Hasse diagrams for the determined subalgebras of $A_{f,K}$. Is f or (f) contained in $A_{f,K}$?

Excercise 328 *For the following combinations of a matrix and a field determine the generalized Jordan decomposition (if it is existing):*

(i) $K := GF(5); M := \begin{pmatrix} 0_K & 0_K & 0_K \\ 1_K & 0_K & 0_K \\ 1_K & 1_K & 1_K \end{pmatrix}$

(ii) $K := GF(2); M := \begin{pmatrix} 1_K & 0_K & 0_K \\ 1_K & 1_K & 0_K \\ 1_K & 1_K & 1_K \end{pmatrix}$

(iii) $K := GF(2)(t); M := \begin{pmatrix} 0_K & 0_K & 0_K \\ 1_K & 1_K & 0_K \\ 0_K & 0_K & t \end{pmatrix}$

(iv) $K := GF(3)(t); M := \begin{pmatrix} t & 0_K & 0_K \\ 1_K & 0_K & 0_K \\ 1_K & 1_K & 1_K \end{pmatrix}$

(v) $K := \mathbb{R}; M := \begin{pmatrix} 0_K & 0_K & 1_K \\ 0_K & 1_K & 0_K \\ 1_K & 0_K & 0_K \end{pmatrix}$

(vi) $K := \mathbb{C}; M := \begin{pmatrix} 0_K & 0_K & 1_K \\ 0_K & 1_K & 0_K \\ 1_K & 0_K & 0_K \end{pmatrix}$

(vii) $K := \mathbb{Q}; M := \begin{pmatrix} 1_K & 1_K & 1_K \\ 1_K & 1_K & 1_K \\ 1_K & 1_K & 1_K \end{pmatrix}$ and

(viii) $K := \mathbb{Q}(i); M := \begin{pmatrix} 1_K & 0_K & 0_K \\ 0_K & 0_K & 0_K \\ i & 1_K & 1_K \end{pmatrix}$.

On what terms is the matrix splittable?

Excercise 329 *Within exercise 328 determine those algebras which are contained in $\Delta_{u,3}$. Only for these algebras compare the generalized Jordan decomposition for an arbitrary element with the decomposition for the element obtained by the decomposition of $\Delta_{u,3}$ into $D(n,3)$ and $rad(\Delta_{u,3})$. Does a radical complement of $\Delta_{u,3}$ exist such that these two decompositions are identically for a fixed arbitrary element or for all elements simultaneously?*

Commutative algebras   205

Excercise 330 *Within example 5 determine the minimum polynomial of an arbitrary element represented in the basis $\{e, r\}$. Which properties are visible using this polynomial?*

Excercise 331 *Within the proof of proposition 15 two statements are included about nilpotent elements. Formulate and prove these statements in details!*

Excercise 332 *Do a research in wikipedia.org for the Chinese-Remainder theorem.*

Excercise 333 *Formulate and prove the Chinese-Remainder theorem used in this chapter (see also exercise 332).*

Excercise 334 *Prove proposition 13 in details.*

Excercise 335 *Prove proposition 14 in details.*

Excercise 336 *Let G be a finite group of order r. If G is Abelian, then $\mid G \mid = \sum_{d \mid r} \mid \{a \mid a \in G, o(a) = d\} \mid$ is valid. Is the opposite statement true, too? In what way does this statement differ for a cyclic group G?*

Excercise 337 *Let G be a finite group of order r and K be a field. On what terms is KG diagonalizable?*

Excercise 338 *Determine the smallest field extension K of $\mathbb{Q}$ such that for all finite Abelian groups G the group algebra KG is diagonalizable. What is the importance of that field extension within algebraic number field theory?*

Excercise 339 *Let G be a finite Abelian group and K be a field such that KG is semisimple. On what terms is the number of irreducible characters of KG exactly $\mid G \mid$?*

Excercise 340 *In this exercise we focus on commutative group algebras based on the following Abelian groups G and fields K:*

(i) $K := \mathbb{C}; G := Z_2$

(ii) $K := \mathbb{C}; G := Z_{49}$

(iii) $K := \mathbb{C}; G := Z_{81}$

(iv) $K := \mathbb{C}; G := Z_{3969}$

(v) $K := \mathbb{C}; G := Z_{7938}$

(vi) $K := \mathbb{R}; G := Z_2$

(vii) $K := \mathbb{R}; G := Z_{49}$

(viii) $K := \mathbb{R}; G := Z_{81}$

(ix) $K := \mathbb{R}; G := Z_{3969}$

(x) $K := \mathbb{R}; G := Z_{7938}$

(xi) $K := \mathbb{Q}(i); G := Z_2$

(xii) $K := \mathbb{Q}(i); G := Z_{49}$

(xiii) $K := \mathbb{Q}(i); G := Z_{81}$

(xiv) $K := \mathbb{Q}(i); G := Z_{3969}$

(xv) $K := \mathbb{Q}(i); G := Z_{7938}$

(xvi) $K := \mathbb{Q}; G := Z_2$

(xvii) $K := \mathbb{Q}; G := Z_{49}$

(xviii) $K := \mathbb{Q}; G := Z_{81}$

(xix) $K := \mathbb{Q}; G := Z_{3969}$ and

(xx) $K := \mathbb{Q}; G := Z_{7938}$.

Determine – if possible – a basis and the dimension for the following K-subalgebras of KG:

(a) $H(KG)$

(b) $D(KG)$

(c) $Sep(KG)$

(d) $VSep(KG)$

(e) $rad(KG)$ and

(f) $ZF(KG)$.

Visualize the results for each example by using Hasse diagrams. In addition, calculate – if possible – the generalized Jordan decomposition of an arbitrary element of KG, decompose KG into fields and calculate the irreducible characters and their numbers.

Excercise 341 *In this exercise we focus on commutative group algebras based on the following Abelian groups G and fields K:*

(i) $K := GF(2); G := Z_4$

Commutative algebras 207

(ii) $K := GF(2); G := Z_2 \times Z_9$

(iii) $K := GF(2); G := Z_2 \times Z_2 \times Z_{5^2} \times Z_7$

(iv) $K := GF(2); G := Z_{2^3} \times Z_{3^3}$

(v) $K := GF(4); G := Z_4$

(vi) $K := GF(4); G := Z_2 \times Z_9$

(vii) $K := GF(4); G := Z_2 \times Z_2 \times Z_{5^2} \times Z_7$

(viii) $K := GF(4); G := Z_{2^3} \times Z_{3^3}$

(ix) $K := GF(3); G := Z_9$

(x) $K := GF(3); G := Z_9 \times Z_4$

(xi) $K := GF(3); G := Z_9 \times Z_5$

(xii) $K := GF(3); G := Z_{17^2} \times Z_{11}$

(xiii) $K := GF(9); G := Z_9$

(xiv) $K := GF(9); G := Z_9 \times Z_4$

(xv) $K := GF(9); G := Z_9 \times Z_5$ and

(xvi) $K := GF(9); G := Z_{17^2} \times Z_{11}$.

Determine – if possible – a basis and the dimension for the following K-subalgebras of KG:

(a) $H(KG)$

(b) $D(KG)$

(c) $Sep(KG)$

(d) $VSep(KG)$

(e) $rad(KG)$ and

(f) $ZF(KG)$.

Visualize the results for each example by using Hasse diagrams. In addition, calculate – if possible – the generalized Jordan decomposition of an arbitrary element of KG, determine the unique radical complement of KG, decompose the radical complement into fields and calculate the irreducible characters and their numbers for the radical complement.

Excercise 342 Prove that minimal polynomials do not change under scalar extensions.

Excercise 343 Let $(K; L)$ a finite field extension and $f, g \in K[t]$. If f is a divisor of g in $L[t]$, then it is a divisor in $K[t]$.

Excercise 344 For the following algebras A determine or describe the intersection of all radical complements and analyze whether the sets $VSep(A)$, $Sep(A)$, $Nil(A)$, $rad(A)$, $D(A)$, $H(A)$ and $ZF(A)$ are subalgebras or subspaces:

(i) $\Delta_{u,n}$ ($n \in \mathbb{N}$, K a field)

(ii) $\Delta_{o,n}$ ($n \in \mathbb{N}$, K a field)

(iii) Solomon-Tits algebra over an arbitrary field

(iv) Solomon algebra in arbitrary dimension over a field of characteristic zero

(v) the real quaternion algebra

(vi) $A(a, b)$

(vii) $K^{n \times n}$ ($n \in \mathbb{N}$, K a field)

(viii) $A^{n \times n}$ ($n \in \mathbb{N}$, K a field, A an associative separable algebra)

(ix) A is commutative associative and finite-dimensional.

(x) A is basic associative and finite-dimensional.

(xi) A is solvable associative and finite-dimensional.

(xii) A is semisimple associative and finite-dimensional.

Excercise 345 Let K be a field, A be a finite-dimensional associative unitary K-algebra, $a \in A$, α resp. γ an algebra monomorphism resp. epimorphism of A and ρ resp. λ the right resp. left regular representation of A. Prove resp. disprove the following statements:

(i) $min_{a,K} = min_{a\rho,K} = min_{a\lambda,K}$

(ii) $min_{a,K} = min_{a\alpha,K}$

(iii) $min_{a,K} = min_{a\gamma,K}$

(iv) $min_{a,K} \mid min_{a\gamma,K}$

(v) $min_{a\gamma,K} \mid min_{a,K}$

Commutative algebras 209

(vi) $char_{a\rho,K} = char_{a\lambda,K}$

(vii) $tr_{a\rho,K} = tr_{a\lambda,K}$

(viii) a is an unit $\iff det(a\rho) \neq 0 \iff det(a\lambda) \neq 0 \iff t$ is no divisor of $min_{a,K}$

(ix) a is a zero divisor $\iff det(a\rho) = 0 \iff det(a\lambda) = 0 \iff t$ is a divisor of $min_{a,K}$

(x) $a \in Nil(A) \implies tr(a\rho) = tr(a\lambda) = 0$

(xi) $tr(a\rho) = tr(a\lambda) = 0 \implies a \in Nil(A)$

(xii) $a \in Nil(A) \implies \exists n \in \mathbb{N} : min_{a,K} = t^n$

(xiii) a semisimple/separable/fully separable/nilpotent/diagonizable/splitting/unit/zero divisor $\implies a\rho$ semisimple/separable/fully separable/nilpotent/diagonizable/splitting/ unit/zero divisor

(xiv) a semisimple/separable/fully separable/nilpotent/diagonizable/splitting/unit/zero divisor $\implies a\lambda$ semisimple/separable/fully separable/nilpotent/diagonizable/splitting/ unit/zero divisor

(xv) Let $A/rad(A)$ be separable. A separable subalgebra S exists such that for all $b \in A$ an element $s \in S$ exists such that $tr(a\rho) = tr(s\rho)$.

(xvi) Let $A/rad(A)$ be separable. A separable subalgebra S exists such that for all $b \in A$ an element $s \in S$ exists such that $tr(a\lambda) = tr(s\lambda)$.

Excercise 346 Let K be a field, A be a finite-dimensional associative unitary K-algebra, ρ resp. λ the right resp. left regular representation of A, $a \in Sep(A)$, $b \in A$ and $(v;n)$ a generalized Jordan decomposition of a with fully-separable part v and nilpotent part n. b is centralizing a if and only if b is centralizing v and n. A subspace of A is $a\rho$- resp. $a\lambda$-invariant if and only if it is $n\rho$- and $v\rho$- resp. $n\lambda$- and $v\lambda$-invariant. a is an unit if and only if v is an unit. In this case $a = v(1 + v^{-1}n)$ is valid. In this multiplicative decomposition v is fully separable and the second factor is unipotent. An unipotent element is of the form $1 + x$ such that x is nilpotent. For an unit switch between the additive and multiplicative decomposition. Do the factors of a multiplicative decomposition commute? Is the decomposition also unique? Are the factors also polynomials of a? Prove that the trace resp. determinant of an unipotent element is $dim_K(A)$ resp. 1.

Excercise 347 Let K be a field, A be a finite-dimensional associative unitary K-algebra, ρ resp. λ the right resp. left regular representation of A, $a \in Sep(A)$, U a $a\rho$- resp. $a\lambda$-invariant subspace of A and $(v;n)$ a generalized Jordan decomposition of a with fully-separable part v and nilpotent part

n. Determine the generalized Jordan decomposition of $a\rho$ resp. $a\lambda$ restricted to U and of the induced endomorphism on A/U.

Excercise 348 Let K be a field, A a finite-dimensional associative unitary K-algebra $k \in K$ and $a, b, c \in Sep(A)$ such that a, b are commuting. Determine a generalized Jordan decomposition for ka, $a + b$, $a \cdot b$ and $a \otimes c$ based on the ones for a, b, c. Is it necessary that the condition $a \circ b = 0$ is valid?

Excercise 349 Let K be a field $f \in K[t]$. Analyze on what terms the factor algebra $K[t]/(f)$ is nilpotent, separable, semisimple, split, diagonizable or reduced?

Excercise 350 Find examples of the generalized quaternion algebras $A(a, b)$ such that some of the elements do not possess a generalized Jordan decomposition.

Excercise 351 Let $K \in \{\mathbb{R}, \mathbb{Q}, \mathbb{C}\}$ or a finite field, $a \in K$, $A := End_K(K^4)$, B the standard basis of K^4 and $\gamma \in A$ defined by the matrix
$$M_B(\gamma) = \begin{pmatrix} 0 & 0 & 0 & -1 \\ a & 1 & 0 & 0 \\ 0 & a & 0 & -2 \\ 0 & 0 & 1 & 0 \end{pmatrix}.$$
Determine a generalized Jordan decomposition for γ. On what terms is γ nilpotent, separable, semisimple, split, diagonizable or fully-separable? On what terms is γ invertible? Apply exercise 346 to this exercise.

Excercise 352 Do a research in the literature and analyze the connection between the Jordan decomposition and the Jordan normal or canonical form!

Excercise 353 Let $K := GF(2)(t)$ and A be the unitary algebra with basis $\{1, a, b, c\}$ defined by the multiplication $a^2 = t1, ab = c, ac = tb, ba = c, b^2 = 0, bc = 0, ca = tb, cb = 0, c^2 = 0$. Prove that A is a associative and commutative K-algebra possessing no radical complement. (Hint: see [40])

Excercise 354 Let A be an associative unitary algebra and $a \in A$ algebraical. Describe in what way $min_{a,K}$ and/or $char_{a,K}$ can be used to determine powers of a (especially if the exponent is of high value).

Excercise 355 Let $K = \mathbb{Q}$, $A := End_K(K^3)$, B the standard basis of K^3 and $\gamma \in A$ defined by the matrix
$$M_B(\gamma) = \begin{pmatrix} 0 & 0 & 2 \\ 1 & 0 & 1 \\ 0 & 1 & -1 \end{pmatrix}.$$
Calculate A^{1000} by using exercise 354.

Commutative algebras   211

Excercise 356 Let A be an associative finite-dimensional unitary algebra and $a, b \in A$. True or false: $min_{ab,K} = min_{ba,K}$.

Excercise 357 Let A be an associative finite-dimensional unitary algebra and $a, b \in A$. True or false: $char_{ab,K} = char_{ba,K}$.

Excercise 358 Let K be a field, A be an associative finite-dimensional unitary algebra, $a \in A$, $f \in K[t]$ such that $f(a) = 0$ (e.g. $min_{a,K}$ or $char_{a,K}$) and $f_1^{r_1} \cdots f_s^{r_s}$ the decomposition of f into irreducible polynomials in $K[t]$. Regard A as $K[t]$-module by using $\alpha = a\rho$ resp. $\alpha = a\lambda$. A is the direct sum of the α-invariant kernels $ker(f_1^{r_1}(\alpha)), \ldots, ker(f_s^{r_s}(\alpha))$. In what can this result be used to determine $min_{a,K}$ or $char_{a,K}$? (Hint: use exercise 324)

Excercise 359 Let A be an associative finite-dimensional unitary algebra and $a \in A$. We call a cyclic if and only if the endomorphism $a\rho$ is cyclic. In this case A is regarded as $K[t]$-module via the action of $a\rho$. Prove that a is cyclic if and only if $a\lambda$ is cyclic. This is only valid if $A = K[a]$ is true. (Hint: The minimal polynomial of a cyclic endomorphism is of degree of the underlying vector space. What is the consequence for the dimension of $K[a\rho]$?)

Excercise 360 Prove that the set of cyclic elements of an associative finite-dimensional unitary algebra needs not be additive or multiplicative closed. Is the unit element cyclic? Is the zero element cyclic? If a is cyclic, then ka is cyclic for every $k \in K$. Let K be a field and G a finite cyclic group. Prove that each generator of G is cyclic within KG. (Hint: use exercise 359)

Excercise 361 Let A be an associative finite-dimensional unitary algebra and $a \in E(A)$. Prove that a^{-1} is a polynomial in a.

Excercise 362 Let V be a finite dimensional K-space, α an endomorphism of V and $V = V_r \leq V_{r-1} \leq \cdots \leq V_1 \leq V_0 = \{0\}$ a sequence of α-invariant subspaces of V. Do a research in the literature (e.g in [37]) to determine a formula or a bound for the calculation of $tr(\alpha)$, $det(\alpha)$, $char_{\alpha,K}$ and $min_{\alpha,K}$ based on the successive factor spaces V_{i+1}/V_i.

Excercise 363 Let A be an associative finite-dimensional unitary algebra, $n := dim_K(A)$ and $a \in E(A)$. Prove that $min_{a,K}(0) \neq 0$ and $char_{a,K}(0) \neq 0$. Now let f be a monic polynomial such that $f(a) = 0$ and $f(0) \neq 0$ are valid. Let $f = t^r + \sum_{i=1}^{r-1} a_i t^i + a_0$. Prove the following statements:

(i) $a^{-1} = -(a_0)^{-1} \sum_{i=1}^{r} a_i a^{i-1}$

(ii) $a^{-1} = \frac{f(t)-a_0}{f(0) \cdot t}(a)$

(iii) $min_{a^{-1},K} = (min_{a,K}(0))^{-1} \cdot t^{deg(min_{a,K})} \cdot min_{a,K}(1/t)$

(iv) $char_{a^{-1},K} = ((-1)^n/(char_{a,K}(0))) \cdot t^n \cdot char_{a,K}(1/t)$

(Hint: see also the exercises within [37]) Apply this exercise to units of $\mathbb{H}$ and to two automorphism of $End_K(K^3)$.

Excercise 364 *Do a research in the literature to find a connection between the trace, the determinant and the characteristical polynomial of an endomorphism.*

Excercise 365 *Let A be an associative finite-dimensional unitary algebra, $n := dim_K(A)$ and $a \in A$. Within this exercise the reader has to link characteristics of a to characteristics of $min_{a,K}$ and to $char_{a\rho,K}$. The same has to be done for $min_{a,K}$ and $char_{a\lambda,K}$, too. The characteristics of a are nilpotent/unipotent/diagonizable/semisimple/fully separable/ being an unit/being a zero divisor/splittable and separable. Find the corresponding characteristics for the mentioned polynomials (e.g. t divides, t divides not, t^n, t^s, $s \leq n$, $(t-1)^n$, $(t-1)^s, s \leq n$, splitting in distinct linear factors, splitting in distinct separable factors, every irreducible factor is separable, every irreducible factor is separable. List the results within a table. Prove all of these characteristics. Furthermore prove that for a semisimple element every $a\rho$-invariant subspace possesses a $a\rho$-invariant complement.*

Excercise 366 *True or false: Two endomorphism possessing the same minimal polynomials are possessing the same characteristical polynomials.*

Excercise 367 *True or false: Two endomorphism possessing the same characteristical polynomials are possessing the same minimal polynomials.*

Excercise 368 *Prove theorem 3 within the article [13]. Use exercise 345. Is it possible to generalize the theorem to algebras possessing a separable factor algebra by its nilradical?*

Excercise 369 *Let A be an associative K-algebra. A derivation of A is a K-linear function $D : A \longrightarrow A$ such that for all $a, b \in A$ the derivation rule $d(ab) = d(a)b + ad(b)$ is valid. Prove that the set $Der(A)$ is a K-Lie subalgebra of $End_K(A)^\circ$.*

Excercise 370 *Let A be an associative K-algebra. For every element $a \in A$ the adjoint representation $ad(a) : A \longrightarrow A$ is defined by $x \mapsto x \circ a := xa - ax$. Prove that $ad(a)$ is a derivation of A and the set $\{ad(a) \mid a \in A\}$ is a Lie ideal of $Der(A)$. These derivations are called inner derivations of A. In what way are inner derivations related to A and the center $Z(A)$ of A? (Hint: see exercise 369)*

Commutative algebras 213

Excercise 371 Let A be an associative unitary K-algebra and $d \in Der(A)$ (see exercise 369). Prove that $d(1) = 0$ and for every $a \in E(A)$ the rule $d(a^{-1}) = -a^{-1}d(a)a^{-1}$ is valid. What is the consequence for an inner derivation of A? Extend this exercise to non-unitary algebras!

Excercise 372 Let A be an associative K-algebra and $d \in Der(A)$ (see exercise 369). If $char(K) = 2$ is valid, then prove that the composition of d with itself is a derivation of A. If $char(K) \neq 2$ is valid and A possesses no zero divisors, then the composition of d with itself is a derivation of A if and only if $d = 0$.

Excercise 373 Determine $Der(A)$ and the set of all inner derivations for the following associative unitary algebras:

(i) $dim_K(A) = 1$

(ii) $dim_K(A) = 2$

(iii) a generalized quaternion algebra as presented in chapter 4

(iv) KC_3 for an arbitrary field K

(v) $\Delta_{u,3}$ for an arbitrary field K.

Excercise 374 Let A be an associative K-algebra, $a \in Sep(A)$ and $(x;y)$ a generalized Jordan decomposition of a. Prove that $(ad(x); ad(y))$ is a generalized Jordan decomposition of $ad(a)$. (Hint: see [58] and [55]) Provide some examples based on exercise 373.

Excercise 375 Let A be an associative K-algebra and d a derivation possessing a generalized Jordan decomposition. True or false: each member of the decomposition is a derivation, too. (Hint: see [58]) Provide some examples based on exercise 373.

Excercise 376 Let K be a field and G a finite cyclic group. The aim of this exercise is to determine the K-subalgebras $H(KG)$, $D(KG)$, $VSep(KG)$, $ZF(KG)$, $rad(KG)$ and $Sep(KG)$. In addition, the generalized Jordan decomposition for the elements of $Sep(KG)$ are to be calculated. Consider – as done in the corresponding section 5.6 of this chapter – the cases 'semisimple' and 'modular'. Apply the results to a cyclic group or prime power order p and finite fields K, L such that $char(K) = p$ and $char(L) \neq p$ are valid.

Excercise 377 In view of exercise 376 is it possible to derive a result by using a direct composition of the cyclic group into cyclic groups of prime power order for different primes concerning the mentioned K-subalgebras? Is this method useful for arbitrary finite Abelian groups?

Excercise 378 Let K be a field and G a finite cyclic group of order n. Prove that KG and $K[t]/(t^n - 1)$ are isomorphic as algebras.

Excercise 379 Let K be a field and G a finite cyclic group. In the articles [19] and [26] the group ring KG is connected to circulant matrices over K. Define circulant matrices over K and prove the mentioned isomorphism. In what way is the isomorphism related to the right and left regular representation of G? Do explicit calculation for the basis elements of cyclic groups of order $2, 3, 5$ and 7!

Excercise 380 Let K be a field and $n \in \mathbb{N}$. Prove that products, sums, multiplications with scalars and inverses of circulant matrices are circulant matrices. Use exercise 379. Do explicit calculation for the basis elements of cyclic groups of order $2, 3, 5$ and 7!

Appendix A

About a theorem of Thorsten Bauer

As mentioned in chapter 3, section 2 the proof of theorem 5.4 in [3] has to be worked out in details. Thus, we finalize the proof and, in addition, we transfer the result to non-unitary associative algebras. At the end we will visualize the argumentation needed in [3] to finalize the proof.

A.1 The proof

The proof needs some preliminary facts.

Lemma 10 *(group of units of the factor algebra by the nilradical) Let K be a field and A a finite-dimensional associative unitary K-algebra. $E(A)/(1_A + rad(A)) = E(A/rad(A))$ is valid.*

Proof. Let $x \in E(A)$. The statement $x(1_A + rad(A)) = x + xrad(A) = x + rad(A) \in E(A/rad(A))$ is true. Let $a + rad(A) \in E(A/rad(A))$. An element $b \in A$ exists such that $ab \in 1_A + rad(A)$ is valid. We deduce $ab \in E(A)$. By using theorem 1.2.1 in [8] we conclude that a is an unit or a divisor of zero of A. Within the first case $a + rad(A) = a(1_A + rad(A)) \in E(A)/(1_A + rad(A))$ is valid. If a is a divisor of zero of A, then an element $0_A \neq c \in A$ exists such that $ca = 0_A$ is true. Hence, $cab = 0_A$ is valid. ab is an unit of A and thus $c = 0_A$ is valid. This statement contradicts the fact $c \neq 0$. ⋄

The lemma is a basic fact. In addition, we need the following deep result of Stuth (see e.g. corollary 5.3.1.2 in [23]).

Theorem 51 *(Stuth) Let D be a skew field. If U is a solvable subnormal subgroup of $E(D)$, then U is central. In particulary, D is a field if $E(D)$ is solvable.* ⋄

We remark that Hua has mentioned in [17] that a skew field is a field if and only if its group of units is solvable. Scott has also proven a generalization of this fact (see e.g. [55]).

Now we finalize the proof presented by Thorsten Bauer.

Theorem 52 *(Bauer) Let K be a field, $\mid K \mid > 3$ and A a finite-dimensional associative unitary K-algebra. If $E(A)$ is solvable, then A is solvable.*

Proof. Let $E(A)$ be solvable. Hence, the factor group $E(A)/(1_A + rad(A))$ is solvable, too. By using lemma 10 we deduce the solvability of $E(A/rad(A))$. The classical result of Wedderburn-Artin about the structure of semisimple associative algebras lets us deduce that finite-dimensional associative K-divisions algebras $D_1, ..., D_r$ and $n_1, ..., n_r \in \mathbb{N}$ exist such that $A/rad(A) \cong_{A_1} \bigoplus_{i=1}^{r} D_i^{n_i \times n_i}$ is valid. We conclude that for all $i \in \underline{r}$ a subalgebra of $A/rad(A)$ exist which is isomorphic to $K^{n_i \times n_i}$. Hence, for all $i \in \underline{r}$ a subgroup of $E(A/rad(A))$ exists which is isomorphic to $GL(n_i, K)$. All of these subgroups are solvable. By using 6.10 in [18] for all $i \in \underline{r}$ the condition $n_i = 1$ is valid. Thus, $A/rad(A)$ is isomorphic to a direct sum of K-division algebras. By using the assumption all group of units of these K-division algebras are solvable, and the theorem of Stuth 51 lets us deduce that they are indeed Abelian. By definition A is solvable.◊

For a comparison we quote the argumentation of T. Bauer in Satz 5.4 in [3]):

> Let $\mid K \mid > 3$ and $E(A)$ solvable. Then $E(A)/(1_A + rad(A))$ is solvable. $A/rad(A)$ is isomorphic to a direct sum of matrix algebras and thus $E(A)/(1_A + rad(A))$ is isomorphic to a direct product of linear groups. The groups $GL(n, K)$ is solvable for a field with $\mid K \mid > 3$ if and only if $n = 1$ is valid (see [18], 6.10). Thus, $A/rad(A)$ is isomorphic to a direct sum of fields.

Hence, the gap in the argumentation of Satz 5.4 in [3] can be closed by using lemma 10 and the theorem of Stuth 51.

As stated before we will now transfer theorem 5.4 in [3] to non-unitary associative algebras.

Theorem 53 *Let K be a field and A a finite-dimensional associative K-algebra. If A is solvable, then $Q(A)$ is solvable, too. If K possesses at least 4 elements then both statements are equivalent.*

Proof. Let A be solvable. By using part (i) of remark 17 we deduce the solvability of (K, A). Theorem 5.4 in [3] lets us conclude that $E(K, A)$ is

solvable. Hence, by using part (v),(c) of remark 17 we deduce the solvability of $Q(K, A)$. $Q(A)$ is isomorphic to a subgroup of $Q(K, A)$ and hence it is solvable, too.
Let $Q(A)$ be solvable and $\mid K \mid > 3$. By using lemma 2.1.2 in [51] we deduce the solvability of $Q(K, A)$. Hence, again by using part (v),(c) of remark 3, $E(K, A)$ is solvable. Theorem 52 implies the solvability of (K, A). A is isomorphic to a K-subalgebra of (K, A) and hence it is solvable, too.◇

By using theorem 21 we deduce a connection between solvable groups, associative and Lie algebras:

Theorem 54 *Let K be a field with at least 4 elements, $char(K) \neq 2$ and A a finite-dimensional associative K-algebra. The following statements are equivalent:*

(i) A is solvable.

(ii) A° is solvable.

(iii) $Q(A)$ is solvable.

If A is unitary, then all three statements are equivalent to the statement that $E(A)$ is solvable.◇

A.2 Exercises

Excercise 381 *Let A be an associative K-algebra and I a K-ideal of A. True or false:*

(i) *If I is nilpotent, then $1+I$ is a nilpotent normal subgroup of $1+rad(A)$ and of $E(A)$.*

(ii) *If I is nilpotent, then I is a nilpotent normal subgroup of $Q(A)$ and $rad(A^\star)$.*

(iii) *If I is nilpotent, then $Q(A)/I$ and $Q(A/I)$ are isomorphic.*

(iv) *If I is nilpotent, then $Q(A)/I = Q(A/I)$ is valid.*

(v) *If I is nilpotent, then $E(A)/(1+I)$ and $E(A/I)$ are isomorphic.*

(vi) *If I is nilpotent, then $E(A)/(1+I) = E(A/I)$ is valid.*

Excercise 382 *Let A be a finite-dimensional associative unitary K-algebra. True or false:*

(i) *If $char(K) = 2$ is valid, then the solvability of A, $E(A)$ and A° are equivalent.*

(ii) If $|K| \leq 3$ is valid, then the solvability of A, $E(A)$ and $A°$ are equivalent.

(iii) If $char(K) \neq 2$ and $|K| \geq 4$ are valid, then the solvability of A, $E(A)$ and $A°$ are equivalent.

Excercise 383 Let A be a finite-dimensional associative K-algebra. True or false:

(i) If $char(K) = 2$, then the solvability of A, $Q(A)$ and $A°$ are equivalent.

(ii) If $|K| \leq 3$, then the solvability of A, $Q(A)$ and $A°$ are equivalent.

(iii) If $char(K) \neq 2$ and $|K| \geq 4$, then the solvability of A, $Q(A)$ and $A°$ are equivalent.

Excercise 384 Let K be a field, $a, b \in K$, $n \in \mathbb{N}$, A, B finite-dimensional associative unitary K-algebras, e a central idempotent of A, M a finite monoid and G a finite group. For the following algebras A analyze on what terms A, $A°$ and $E(A)$ resp. $Q(A)$ are solvable and determine their class of solvability:

(i) $\Delta_{u,n}$

(ii) $\Delta_{o,n}$

(iii) KM for M being commutative and idempotent

(iv) KG for G being Abelian

(v) KG for $G = Q_8$ and $K := \mathbb{Q}$

(vi) KG for $G = D_8$ and $K := \mathbb{R}$

(vii) KG for $G = SD_8$ and $K := \mathbb{C}$

(viii) $K^{n \times n}$

(ix) $\mathbb{H}$

(x) eAe for A being solvable (see exercise 173)

(xi) zero extension of A for A being separable and solvable (see exercise 174)

(xii) $A(a, b, K)$ for $char(K) = 2$

(xiii) $A(a, b, K)$ for $char(K) \neq 2$

Appendix B

Proof of the Wedderburn-Malcev theorem for associative unitary algebras

Within this chapter we want to prove the theorem of Wedderburn-Malcev for unitary algebras:

Let K be a field and A a finite-dimensional associative unitary K-algebra. If $A/rad(A)$ is separable, then the following statements are valid:

(i) *A possesses a radical complement.*

(ii) *The nilpotent normal subgroup $1_A + rad(A)$ of $E(A)$ acts transitive per conjugation on the set of all radical complements of A.*

We use the proofs within [8], [9] and [10]. The analysis of the proof within [10] leads to a generalized theorem for the conjugacy part which is proven here:

Let K be a field, A a finite-dimensional associative unitary K-algebra, S a radical complement (not necessarily separable) and T an unital separable K-subalgebra of A. T can be conjugated into S by a suitable unit $1 + r$, $r \in rad(A)$.

B.1 The existence part

B.1.1 The case of a zero nilradical - by using cohomology of algebras

The proof is based on the proof stated within [35]. Let $rad(A)^2 = 0$ and π the natural algebra epimorphism from A onto $A/rad(A)$ defined by $a \mapsto a + rad(A)$. We use a K-linear function $\kappa : A/rad(A) \longrightarrow A$ such that $\kappa\pi = id_{A/rad(A)}$ is valid. Here, we use only standard linear algebra. The aim of the proof is to enrich the linear function κ such that the enrichment is an algebra homomorphism. For this, we use the concept of factor sets described within the section of separable algebras. We define

$$\Psi : A/rad(A) \times A/rad(A) \longrightarrow rad(A) \text{ by } \Psi(x,y) := \kappa(xy) - \kappa(x) \cdot \kappa(y).$$

This function measures the deviation of κ from being an algebra homomorphism. Let $x, y \in A/rad(A)$. Indeed,

$$\begin{aligned} \Psi(x,y)\pi &= \\ \kappa(xy)\pi - (\kappa(x)\kappa(y))\pi &= \\ xy - \kappa(x)\pi\,\kappa(y)\pi &= \\ xy - xy &= \\ 0, \end{aligned}$$

and thus $\Psi(x,y) \in ker(\pi) = rad(A)$.

The next step is to establish a bimodule structure on $rad(A)$ induced by κ and by using the separable algebra $A/rad(A)$. We define

$$u.x := u\kappa(x) \text{ and } x.u := \kappa(x)u$$

for all $x \in A/rad(A)$ and $u \in rad(A)$. Most of the axioms are straightforward to prove because of the linearity of κ and the associativity of A. For the right-module laws only $(xy).u = x.(y.u)$ $(x, y \in A/rad(A), u \in rad(A))$ is not straightforward to prove. Let $x, y \in A/rad(A)$ and $u \in rad(A)$. We calculate

$$\begin{aligned} (xy).u - x.(y.u) &= \\ \kappa(xy)u - \kappa(x)(\kappa(y)u) &= \\ \Psi(x,y)u \in rad(A)^2 &= \\ 0. \end{aligned}$$

The argumentation for the left module laws are identical by using $rad(A)^2 = 0$. Therefor, we omit this calculation here.

Proof of the Wedderburn-Malcev theorem for associative unitary algebras

The next step is to prove that Ψ is a factor set. Straightforward to check is that Ψ is bi-linear (because κ is linear). Thus, we check the factor set rule only: Let $a, b, c \in A/rad(A)$. We calculate:

$$\begin{aligned}
a.\Psi(b,c) - \Psi(ab,c) + \Psi(a,bc) - \Psi(a,b).c &= \\
\kappa(a)(\kappa(bc) - \kappa(b)\kappa(c)) &- \\
(\kappa(abc) - \kappa(ab)\kappa(c)) &+ \\
\kappa(abc) - \kappa(a)\kappa(bc) - (\kappa(ab) &- \\
\kappa(a)\kappa(b))\kappa(c) &= \\
\kappa(a)\kappa(bc) - \kappa(a)\kappa(b)\kappa(c) - \kappa(abc) &+ \\
\kappa(ab)\kappa(c) + \kappa(abc) - \kappa(a)\kappa(bc) &- \\
\kappa(ab)\kappa(c) + \kappa(a)\kappa(b)\kappa(c) &= \\
0.
\end{aligned}$$

The algebra $A/rad(A)$ is separable, and by using theorem 7 we derive that Ψ is a split factor set. Thus, a linear function $\alpha : A/rad(A) \longrightarrow rad(A)$ exists such that for all $x, y \in A/rad(A)$ the identity

$$\Psi(x,y) = x.\alpha(y) - \alpha(xy) + \alpha(y).y$$

is valid. Based on α we define $\Phi := \kappa + \alpha$ which is a linear function from $A/rad(A)$ to A. Now we have enriched κ by α and we want to prove in the last step that Φ is an algebra monomorphism and its image is a radical complement in A. For this, we have to prove that it is a monomorphism preserving the unit element. From this, we derive that the image is a sub-algebra isomorphic to $A/rad(A)$ which is indeed a radical complement.

Let $x, y \in A/rad(A)$. We calculate based on the definition of Φ and the split property of Φ:

$$\begin{aligned}
\Phi(xy) - \Phi(x)\Phi(y) &= \\
\kappa(xy) + \alpha(xy) - (\kappa(x) + \alpha(x))(\kappa(y)\alpha(y)) &= \\
\kappa(xy) + \alpha(xy) - \kappa(x)\kappa(y) + \kappa(x)\alpha(y) - \alpha(x)\kappa(y) - \alpha(x)\alpha(y) &= \\
\Phi(x,y) - \Phi(x,y) - \alpha(x)\alpha(y) \in rad(A)^2 &= \\
0.
\end{aligned}$$

In addition, we derive from $Im(\alpha) \subseteq rad(A) = ker(\pi)$ the condition

$$\Phi\pi = (\kappa + \alpha)\pi = \kappa\pi = id_{A/rad(A)}.$$

We conclude that Φ is injective. Finally, $\Phi(1) = \Phi(1^2) = \Phi(1)^2$ is valid, and thus

$$\begin{aligned}
\Phi(1 + rad(A)) - 1 &= \\
\kappa(1 + rad(A)) + \alpha(1 + rad(A)) - 1 &= \\
\alpha(1 + rad(A)) \in rad(A)
\end{aligned}$$

is true. Therefor,

$$
\begin{aligned}
0 &= \\
(\Phi(1+rad(A))-1)^2 &= \\
\Phi(1+rad(A))^2 - 2\Phi(1+rad(A)) + 1 &= \\
1 - \Phi(1+rad(A)) &
\end{aligned}
$$

is valid. We derive that Ψ preserves the unit element.⋄

B.1.2 The induction argument

Because of subsection B.1.1 we can assume that $rad(A)^2$ is not zero. We use the section B.1.1 and proceed by an induction argument based on the nilpotency class of $rad(A)$ as mentioned within the textbook [8] within chapter 6, page 109. Let $B := A/rad(A)^2$. It is well-known that $rad(B) = rad(A)/rad(A)^2$ is valid. Hence, $rad(B)$ is a zero algebra and $B/rad(B)$ is isomorphic to $A/rad(A)$. Thus, B is a finite-dimensional associative unitary algebra possessing a separable factor algebra by its zero-nilradical. We can apply the previous subsection B.1.1 for the algebra B. Let T be a separable subalgebra of B such that $T \oplus rad(B) = B$ is valid. By using the algebra homomorphism theorem a subalgebra S of A exists such that $S/(rad(A)^2) = T$ is true. Hence, we deduce $S/(rad(A)^2) \oplus rad(A)/(rad(A)^2) = A/(rad(A)^2)$. $rad(A)^2$ is a nilpotent ideal of A such that $S/(rad(A)^2)$ is separable. We conclude $rad(S) = rad(A)^2$. In addition, $S/rad(S)$ is isomorphic to T which is separable. A is unitary, and thus B is unitary, too. We use remark 4 to deduce that T is unital within B. S is the epimorphic pre-image of the natural epimorphism from A to $A/(rad(A)^2)$. Thus, $S = \{a \in A \mid a+rad(A)^2 \in T\}$. T is unital, and thus $1_A + rad(A)^2 \in T$. We deduce $1_A \in S$. We can apply induction on S because the nilpotency class of $rad(A)^2$ is smaller than the one of $rad(A)$. Thus, a subalgebra C exists such that $S = C \oplus rad(A)^2$. In addition, $S/(rad(A)^2) \oplus rad(A)/(rad(A)^2) = A/(rad(A)^2)$ is valid. We prove $A = C \oplus rad(A)$. C is isomorphic to $S/(rad(A)^2)$ which is T. The algebra T is isomorphic to $B/rad(B)$ which is isomorphic to $A/rad(A)$.⋄

B.2 The conjugacy part

B.2.1 The case of a zero nilradical - by using cohomology of algebras

This part is proven as done in [10] by using derivations and bimodules. Let S be a radical complement and T an unital separable subalgebra of A. In addition, let $rad(A)^2 = 0$. Recall, that every derivation from the separable algebra T into a (T,T)-bimodule M is an inner derivation based on theorem 4: $Der(T,M) = Inder(T,M)$. Because of the decomposition

$A = S \oplus rad(A)$ for every element $t \in T$ an unique element $f(t) \in S$ and an unique element $g(t) \in rad(A)$ exists such that $t = f(t) + g(t)$ is valid. We analyze the functions f, g in more details now.

Let $t, \hat{t} \in T$. Then $t = f(t) + g(t)$ and $\hat{t} = f(\hat{t}) + g(\hat{t})$ are valid. We calculate

$$t \cdot \hat{t} = f(t)f(\hat{t}) + f(t)g(\hat{t}) + g(t)f(\hat{t}) + g(t)g(\hat{t}).$$

Because $rad(A)$ is an ideal of A we derive

$$f(t\hat{t}) = f(t)f(\hat{t}) \text{ and } g(t\hat{t}) = f(t)g(\hat{t}) + g(t)f(\hat{t}) + g(t)g(\hat{t}).$$

In addition, $rad(A)^2 = 0$, and thus

$$g(t\hat{t}) = f(t)g(\hat{t}) + g(t)f(\hat{t})$$

is valid. Straightforward to prove are that f, g are K-linear and $f(1) = 1$. Therefore, f is an algebra homomorphism between T and S. The radical $rad(A)$ is a bimodule under the associative algebra multiplication of A, and hence also under the multiplication of T. The corresponding bimodule representations are the right-multiplication

$$\rho : T \longrightarrow End(rad(A)), a \mapsto (a\rho)$$

and the left-multiplication

$$\lambda : T \longrightarrow End(rad(A)), a \mapsto (a\lambda).$$

f is an algebra homomorphism. Hence, $f\rho$ and $f\lambda$ are still valid as bimodule operations. As a consequence, $rad(A)$ is a (T,T)-bimodule with respect to the operations

$$t.r := f(t)r \text{ and } r.t := rf(t)$$

for all $r \in rad(A)$ and $t \in T$. In view of this operation we derive $g \in Der(T, rad(A))$. T is separable, and thus g is an inner derivation: $g \in Inder(T, rad(A)$. We analyze the consequence of g being an inner derivation.

Let $r \in rad(A)$ such that for all $t \in T$ the equation

$$g(t) = ad(t) = t.r - r.t = f(t)r - rf(t)$$

is valid. We calculate

$$t = g(t) + f(t) = f(t)r - rf(t) + f(t) = f(t)(1+r) - rf(t).$$

Because of $rad(A)^2 = 0$ the element $1-r$ is the inverse element of $1+r$. Thus, we derive (again by using $rad(A)^2 = 0$ and thus $r^2 = 0$ and $(rf(t))r = 0$)

$$(1+r)t(1-r) =$$
$$(1+r)f(t) - (1+r)rf(t)(1-r) =$$
$$(1+r)f(t) - rf(t)(1-r) =$$
$$f(t) \in S.$$

The proof is finished.◊

B.2.2 The induction argument

Because of subsection B.2.1 we can assume that $rad(A)^2$ is not zero. We use the section B.2.2 and proceed by an induction argument based on the nilpotency class of $rad(A)$ similar to the proof stated within [10]. Let $B := A/rad(A)^2$ and S a radical complement and T a separable subalgebra of A. It is well-known that $rad(B) = rad(A)/rad(A)^2$ is valid. Hence, $rad(B)$ is a zero algebra and $B/rad(B)$ is isomorphic to $A/rad(A)$. Thus, B is a finite-dimensional associative unitary algebra possessing a zero-nilradical. We can apply the previous subsection B.2.2 for the algebra B. $(S \oplus rad(A)^2)/(rad(A)^2)$ is a radical complement of B isomorphic to S (because $S \cap rad(A)^2$ is a nilpotent ideal of the semisimple subalgebra S and hence must be zero) and $(T \oplus rad(A)^2)/(rad(A)^2)$ is a separable subalgebra of B (because of the same argumentation as for S). Hence, an element $r \in rad(A)$ exists such

$$((T \oplus rad(A)^2)/(rad(A)^2))^{(1+r)+rad(A)^2} \subseteq (T \oplus rad(A)^2)/(rad(A)^2).$$

We deduce $T^{1+r} \subseteq S \oplus rad(A)^2$. Let $X := S \oplus rad(A)^2$. We use remark 4 to deduce that S is unital within A. Therefore, X is a finite-dimensional associative unitary K-algebra possessing the radical $rad(A)^2$ and the radical complement S. T^{1+r} is contained in X and is isomorphic to T. We conclude that T^{1+r} is separable. By an induction argument $(cl(rad(A)^2) < cl(rad(A)))$ an element $x \in rad(A)^2 \leq rad(A)$ exists such that $(T^{1+r})^{1+x} \subseteq T$. $rad(A)$ is a subalgebra, and thus $r + x + rx \in rad(A)$ is valid. We conclude that T can be conjugated into S by using the element $1 + (r + x + rx) \in 1 + rad(A)$.◊

B.3 Open-ended questions and exercises

Open-ended question 4 *(i) Let K be a field, A a finite-dimensional associative unitary K-algebra, S a radical complement (not necessarily separable) and T, R two maximal separable K-subalgebras of A. Are T and R conjugated in A? By using the generalized version of the conjugacy part of the Wedderburn-Malcev theorem both subalgebras can* be conjugated into S. The question is whether two maximal separable subalgebras of a semisimple algebra are conjugated. We have already proven that within a commutative algebra A only one maximal separable subalgebra exists: $VSep(A)$, the set of all separable elements.

Excercise 385 *This exercise is taken from the article [10] of R. Farnsteiner. Let p be a prime number and E be a purely inseparable field extension of exponent 1, $E := K(\alpha)$, $a := \alpha^p \in K$ and $\alpha \notin K$. We define a function $d : E \longrightarrow E$ by $d(\alpha^i) := i\alpha^{-1}$ for all $1 \leq i \leq p-1$. Prove that d is a derivation. We define based on d the function*

$$f : E \times E \longrightarrow E \text{ by } f(a,b) := \sum_{r=1}^{p} \frac{1}{p}\binom{p}{r} d^r(a) d^{p-r}(b).$$

Prove that f is a factor set which does not split.

Excercise 386 *Within subsection B.2.1 prove that $rad(A)$ is a (T,T) bimodule under the action induced by f and that g is a derivation if and only if $rad(A)^2 = 0$.*

Excercise 387 *Transfer the generalized statement of the Wedderburn-Malcev theorem to non-unitary algebras!*

Excercise 388 *Define purely inseparable field extensions of arbitrary exponent.*

Excercise 389 *Let $e \in \mathbb{N}$ and p a prime number. True or false: The field extension $(GF(p)(t); GF(t^{p^e}))$ is a purely inseparable field extension of exponent e.*

Excercise 390 *This exercise is contained as example within the article of R. Farnsteiner (see [10], examples on page 3). Let p be a prime number and E be a purely inseparable field extension of exponent 1, $E := K(\alpha)$, $a := \alpha^p \in K$ and $\alpha \notin K$. We focus on the K-algebra $A := E \otimes_K E$ and the multiplication $\mu : A \longrightarrow A, a \otimes b \mapsto ab$. Prove the following statements:*

(i) μ is a K-algebra homomorphism.

(ii) The image of μ is E.

(iii) The kernel of μ is $\langle x \otimes 1 - 1 \otimes x \mid x \in E \rangle_K$.

(iv) E is a semisimple, but non-separable K-algebra.

(v) $(x \otimes 1 - 1 \otimes x)^p = 0$ for all $x \in E$.

(vi) $\langle x \otimes 1 - 1 \otimes x \mid x \in E \rangle_K$ is a nilpotent ideal.

(vii) $\langle x \otimes 1 - 1 \otimes x \mid x \in E \rangle_K$ is the nilradical of A.

(viii) For all $a \in A$ the elements $a - (a\mu \otimes 1)$ and $a - (1 \otimes a\mu)$ are contained in the kernel of μ.

(ix) A is commutative as K-algebra.

(x) $1 \otimes E$ and $E \otimes 1$ are non-conjugated radical complements of A.

Excercise 391 *Prove the induction argument of subsection B.2.2 again by starting the proof dual with $rad(A)^{cl(rad(A))-1}$ instead of $rad(A)^2$.*

Excercise 392 *Prove the induction argument of subsection B.1.2 again by starting the proof dual with $rad(A)^{cl(rad(A))-1}$ instead of $rad(A)^2$.*

Excercise 393 *Let K be a field, A, B finite-dimensional associative unitary K-algebras, G a finite group, I an K-ideal of A, T a K-subalgebra of A, e an idempotent of A and $n \in \mathbb{N}$. True or false:*

(i) *If A is commutative, then $Der(A) = Inder(A)$ is valid.*

(ii) *If A is solvable, then $Der(A) = Inder(A)$ is true.*

(iii) *If A is separable, then $Der(A) = Inder(A)$ is valid.*

(iv) *If A is separable, then $Der(A^{n \times n}) = Inder(A^{n \times n})$ is valid.*

(v) *If A is semisimple, then $Der(A) = Inder(A)$ is true.*

(vi) *If A is simple, then $Der(A) = Inder(A)$ is true.*

(vii) *If A is a field, then $Der(A) = Inder(A)$ is true.*

(viii) *If A is a division algebra, then $Der(A) = Inder(A)$ is true.*

(ix) *If A is a generalized quaternion algebra, then $Der(A) = Inder(A)$ is true.*

(x) *If $A = KG$ such that KG is semisimple, then $Der(A) = Inder(A)$ is true.*

(xi) *If $A = KG$ such that KG is not semisimple, then $Der(A) = Inder(A)$ is true.*

(xii) *If A is the algebra of strict upper triangular matrices of $K^{n \times n}$, then $Der(A) = Inder(A)$ is true.*

(xiii) *If A is the algebra of strict lower triangular matrices of $K^{n \times n}$, then $Der(A) = Inder(A)$ is true.*

Proof of the Wedderburn-Malcev theorem for associative unitary algebras 227

(xiv) If A is the algebra of diagonal matrices of $K^{n \times n}$, then $Der(A) = Inder(A)$ is true.

(xv) If A is the algebra of upper triangular matrices of $K^{n \times n}$, then $Der(A) = Inder(A)$ is true.

(xvi) If A is the algebra of lower triangular matrices of $K^{n \times n}$, then $Der(A) = Inder(A)$ is true.

(xvii) $Der(K^n) = Inder(K^n)$

(xviii) $Der(A^n) = Inder(A^n)$ if $Der(A) = Inder(A)$ is valid

(xix) If $Der(A) = Inder(A)$ is valid, then $Der(A^{op}) = Inder(A^{op})$ is true.

(xx) If A is a central-simple algebra, then $Der(A) = Inder(A)$ is true.

(xxi) If A is a central-simple division algebra, then $Der(A) = Inder(A)$ is true.

(xxii) If A and B are isomorphic and $Der(A) = Inder(A)$ is valid, then $Der(B) = Inder(B)$ is true.

(xxiii) If A and B are anti-isomorphic and $Der(A) = Inder(A)$ is valid, then $Der(B) = Inder(B)$ is true.

(xxiv) If $Der(A) = Inder(A)$ is valid, then $Der(T) = Inder(T)$ is true.

(xxv) If $Der(A) = Inder(A)$ is valid, then $Der(I) = Inder(I)$ is true.

(xxvi) If $Der(A) = Inder(A)$ is valid, then $Der(A/I) = Inder(A/I)$ is true.

(xxvii) If $Der(A) = Inder(A)$ is valid, then $Der(A^{n \times n}) = Inder(A^{n \times n})$ is true.

(xxviii) If $Der(A) = Inder(A)$ and $Der(B) = Inder(B)$ are valid, then $Der(A \times B) = Inder(A \times B)$ is true.

(xxix) If $Der(A) = Inder(A)$ and $Der(B) = Inder(B)$ are valid, then $Der(A \otimes B) = Inder(A \otimes B)$ is true.

(xxx) If $Der(A) = Inder(A)$ is valid, then $Der(eAe) = Inder(eAe)$ is true.

(xxxi) $Der(eAe) = eDer(A)e$

(xxxii) $Inder(eAe) = eInder(A)e$

(xxxiii) If $Der(A) = Inder(A)$ is valid and e is central, then $Der(eAe) = Inder(eAe)$ is true.

(xxxiv) If $Der(A) = Inder(A)$ is valid, then $Der(Z) = Inder(Z)$ is true for a zero-extension Z of A.

Appendix C

Proof of Taft's theorem for associative unitary algebras

Let K be a field, A a finite-dimensional associative unitary algebra possessing a separable factor algebra by its nilradical and G a finite group acting as anti- or automorphism on A such that $char(K)$ does not divide $|G|$. The auto- resp. anti-automorphism acting by

$$a.g := ag \text{ resp. } a.g := -(ag)$$

for all $a \in A$ and $g \in G$. So, A is a right G-module. We want to prove the existing part of Taft's theorem:

A possesses a G-invariant radical complement.

In addition, we want to prove the uniqueness theorem of Taft's theorem:

Let S, T be two G-invariant radical complements of A and $char(K) \neq 2$. S, T are G-orthogonal conjugated by an element $1 + r$, $r \in rad(A)$.

By analyzing the proofs presented by Taft we are able to enhance the conjugacy part without assuming that $char(K) \neq 2$ is valid:

Let S, T be two G-invariant radical complements of A. S, T are G-orthogonal conjugated by an element $1 + r$, $r \in rad(A)$.

C.1 The existence part

This section is dedicated to the proof of the existence part of Taft's theorem based on Taft's presentations within [41], [42], [43], [44], [45] and [46].

C.1.1 The case of a zero nilradical - cohomology of groups

We focus on the argumentation stated in the cohomology approach by Taft within [41] and [42] and begin this subsection with some background from group cohomology.

If M is a right G-module (with respect to a field K), then we define a 1-cocycle to be a map

$$d : G \longrightarrow M$$

such that for all $g, h \in G$ the rule

$$d(gh) = d(g).h + d(h)$$

is valid. Special 1-cocycles are 1-coboundaries which are functions

$$d_m : G \longrightarrow M$$

based on an element $m \in M$ such that for all $g \in G$ the rule

$$d_m(g) = mg - m$$

is valid. Both sets of functions form groups, and the set of 1-coboundaries is a normal subgroup of the group of 1-cocycles. The factor group is called the first cohomology group of G with respect to M. For more details the reader may study e.g. the presentations within [48] and [18]. If the characteristic of K does not divide the order of G, then the first cohomology group is zero (see e.g. 4.5 in [48]). In some works 1-cocycles are also called derivations and 1-coboundaries are also called inner derivations of G into M.

Within the argumentation we need the following rules which are straightforward to verify:

(i) If $rad(A)^2 = 0$, then for all $r \in rad(A)$ the identity $(1+r)^{-1} = 1 - r$ is valid.

(ii) If $rad(A)^2 = 0$, then for all $r, s \in rad(A)$ the identity $\kappa_{1+r}\kappa_{1+s} = \kappa_{1+r+s}$ is valid.

(iii) Let $\alpha \in Aut(A)$ and $a \in E(A)$. The identity $\alpha^{-1}\kappa_a\alpha = \kappa_{a\alpha}$ is valid.

(iv) Let $\alpha \in Ant(A)$ and $a \in E(A)$. The identity $\alpha^{-1}\kappa_a\alpha = \kappa_{(a\alpha)^{-1}}$ is valid.

(v) If $rad(A)^2 = 0$ and $r \in rad(A)$ a G-skew element, then $1 - r$ is G-orthogonal.

Proof of Taft's theorem for associative unitary algebras

Now we turn to the proof within the case $rad(A)^2 = 0$. Let $A = rad(A) \oplus S$ with separable radical complement S based on the theorem of Wedderburn-Malcev. For all $g \in G$ the subalgebra $S.g$ is also a radical complement. Again, by using the theorem of Wedderburn-Malcev an element $z(g) \in rad(A)$ exists such that

$$S.g = S\kappa_{1+z(g)}$$

is valid. The element $z(g) \in rad(A)$ is not unique, but we prove that is unique modulo $rad(A) \cap Z(A)$. Let $y(g) \in rad(A)$ such that

$$S\kappa_{1+z(g)} = S\kappa_{1+y(g)}$$

is true. We deduce that

$$(1+z(g))(1+y(g))^{-1} \in N_A(S).$$

As already proven in remark 12 we derive

$$(1+z(g))(1+y(g))^{-1} \in C_A(S).$$

Now we use the rules C.1.1 and calculate:

$$(1+z(g))(1+y(g))^{-1} = (1+z(g))(1-y(g)) = 1+z(g)-y(g) \in C_A(S).$$

Thus, $z(g) - y(g) \in C_A(S)$ is valid. $rad(A)^2 = 0$ and we derive $z(g) - y(g) \in C_A(rad(A))$. Hence, $z(g) - y(g) \in Z(A)$ as desired. We focus on the function

$$\hat{z} : G \longrightarrow rad(A)/(rad(A) \cap Z(A)).$$

The space $rad(A) \cap Z(A)$ is invariant under every anti- and automorphism of A, and thus it is G-invariant and $rad(A) \cap Z(A)$ is a G-submodule of A. As a consequence $rad(A)/(rad(A) \cap Z(A))$ is a G-module, too. We prove that $\hat{z}$ is a derivation of G into $rad(A)/(rad(A) \cap Z(A))$. Let $g, h \in G$. We calculate:

$$\begin{aligned} S^{1+z(gh)} &= \\ S.(gh) &= \\ (S.g).h &= \\ (S\kappa_{1+z(g)}).h &= \\ (S.h.h^{-1})\kappa_{1+z(g)}.h &= \\ (S\kappa_{1+z(h)}.h^{-1})\kappa_{1+z(g)}.h. \end{aligned}$$

Thus,

$$S^{1+z(gh)} = (S\kappa_{1+z(h)}.h^{-1})\kappa_{1+z(g)}.h.$$

Now we apply the rules mentioned before. If g is a automorphism, then

$$(S\kappa_{1+z(h)}.h^{-1})\kappa_{1+z(g)}.h =$$
$$S\kappa_{1+z(h)}\kappa_{(1+z(g))h} =$$
$$S\kappa_{1+z(h)}\kappa_{1+z(g)h} =$$
$$S\kappa_{1+z(h)+z(g)h}.$$

If g is an anti-automorphism, then g is action by $a.g := -(ag)$ for all $a \in A$. Thus,

$$(S\kappa_{1+z(h)}.h^{-1})\kappa_{1+z(g)}.h =$$
$$S\kappa_{1+z(h)}\kappa_{((1+z(g))h)^{-1}} =$$
$$S\kappa_{1+z(h)})\kappa_{1-z(g).h} =$$
$$S\kappa_{1+z(h)+z(g)h}.$$

In both cases we derive, that $\hat{z}$ is a 1-cocycle. By our assumption the first cohomology group of G for the G-module $rad(A)/(rad(A) \cap Z(A))$. Thus, $\hat{z}$ is a 1-coboundary and an element $x \in rad(A)$ exists such that $\hat{z}(g) = (x + (rad(A) \cap Z(A))).g - x + (rad(A) \cap Z(A))$. Thus, we can use $z(g) = x - x.g$ and we prove that $S\kappa_{1+x}$ is a G-invariant radical complement. Let g be an automorphism. We calculate based on the rules mentioned before and the 1-coboundary property:

$$(S\kappa_{1+x}).g =$$
$$(S.g.g^{-1})\kappa_{1+x}.g =$$
$$S\kappa_{1+z(g)}\kappa_{1+xg} =$$
$$S\kappa_{1+z(g)+xg} =$$
$$S\kappa_{1+x}.$$

Let g be an anti-automorphism. Again recall, that g acts with its negative on A. We calculate

$$(S\kappa_{1+x}).g =$$
$$(S.g.g^{-1})\kappa_{1+x}.g =$$
$$S\kappa_{1+z(g)}\kappa_{(1-xg)^{-1}} =$$
$$S\kappa_{1+z(g)}\kappa_{1+xg} =$$
$$S\kappa_{1+z(g)+xg} =$$
$$S\kappa_{1+x}.$$

This calculation finishes the proof.◊

C.1.2 The case of a zero nilradical - cohomology of algebras

Within this section we present an alternative proof based on cohomology of algebras likewise done in the previous subsection C.1.1. We proceed as done in subsection C.1.1 up to the topic that $\hat{z}$ is a 1-cocycle from G into the G-module $rad(A)/(rad(A) \cap Z(A))$. $\hat{z}$ can be extended K-linear from G to KG. We want to equip $rad(A)/(rad(A) \cap Z(A))$ with a KG-bimodule structure. From right hand side there is already the structure induced by G. From left we use the trivial module structure $g.a := a$ for all $g \in G$ and $a \in rad(A)/(rad(A) \cap Z(A))$. Thus, $rad(A)/(rad(A) \cap Z(A))$ is a KG-bimodule and recall that under the assumption KG is separable as associative algebra (see theorem 2). It is straightforward to prove the linearization of $\hat{z}$ is a bimodule derivation of KG into $rad(A)/(rad(A) \cap Z(A))$ (because it possesses already the derivation property on G). We apply theorem 4 to deduce that this derivation is an inner derivation. It is again straightforward to prove that the restriction to G is a 1-coboundary of G (because it possesses this property on whole KG and G acts trivial from left hand side). Now the proof continues as done within subsection C.1.1. With this variant we can use algebra cohomology as already done within the Wedderburn-Malcev theorem and avoid group cohomology to be used.◇

C.1.3 The case of a zero nilradical - derivations of algebra

Within this section we present an alternative proof related to the one presented by Taft within [43]. Let T be a radical complement of A which exists by using the Wedderburn-Malcev theorem. We derive that $A = T \oplus rad(A)$ is valid. Furthermore, we assume $rad(A)^2 = 0$. At first we need some preliminary remarks for KG-modules as presented within [30], Proposition 7.3, which are also used for proving the theorem of Maschke. Let $\Pi : A \longrightarrow rad(A)$ be the projection from A to $rad(A)$. Recall, that for all $g \in G$ the endomorphism Π^g is defined by $g^{-1}\Pi g$. If we define

$$\Pi^G := \sum_{g \in G} \Pi^g,$$

then

$$T^G := ker(\Pi^G)$$

is a K-space complement of $rad(A)$ in A which is also G-invariant. For the invariance and the complementary result we need the assumptions that $char(K)$ does not divide $|G|$.

We need to show that T^G is an unital subalgebra of A. For this, we will use again the concept of derivation. We want to prove that Π^G is a derivation of A into a bimodule. By using examples 2 we conclude that the kernel of Π^G is an unital subalgebra. The bimodule is the nilradical as proven within the conjugacy part of the Wedderburn-Malcev theorem with subsection B.2.1: the projection to $rad(A)$ is g and the projection to T is f within that proof. Thus, $rad(A)$ is a (A,A)-bimodule by the operation induced by the homomorphism

$$f\colon r.a := rf(a) \text{ and } a.r := f(a)r$$

for all $a \in A$ and $r \in rad(A)$. Because of $rad(A)^2 = 0$ the projection is a derivation of A into $rad(A)$ with respect to this bimodule structure: $\Pi \in Der(A, rad(A))$. Hence, for all $g \in G$ – as presented in the examples 2 – the map Π^g is also a derivation of A into $rad(A)$. Thus, – again using the examples 2 – the sum of the derivation $\Pi^g, g \in G$ as a derivation of A into $rad(A)$.◇

C.1.4 The induction argument

Because of subsection C.1.1 we can assume that $rad(A)^2$ is not zero. We use the sections C.1.1, C.1.3 and C.1.2 and proceed by an induction argument based on the nilpotency class of $rad(A)$. The proof stated here is based on the one presented by Taft in [43]. G acts as anti- or automorphism on A. Therefore, the nilradical of A is invariant under the action of G. We conclude that all powers of $rad(A)$ are G-invariant. Let $B := A/rad(A)^2$. Then G acts also on B as anti- or automorphism by

$$(a + rad(A)^2)g := (a.g) + rad(A)^2$$

for all $a \in A$ and $g \in G$ (It is straightforward to prove that this action define homo- resp. anti-homomorphism.) By using the finite dimension we need to prove the injectivity. Let $a \in A$ such that $a.g + rad(A)^2 = 0 + rad(A)^2$. We conclude $a.g \in rad(A)^2$. $rad(A)^2$ is G-invariant and we deduce $a = (a.g).g^{-1} \in (rad(A)^2).g^{-1} \leq rad(A)^2$.) It is well-known that $rad(B) = rad(A)/rad(A)^2$ is valid. Hence, $rad(B)$ is a zero algebra and $B/rad(B)$ is isomorphic to $A/rad(A)$. Thus, B is a finite-dimensional associative unitary algebra possessing a separable factor algebra by its zero-nilradical. We can apply the previous subsection C.1.1 for the algebra B. Let T be a separable and G-invariant subalgebra of B such that $T \oplus rad(B) = B$ is valid. By using the algebra homomorphism theorem a subalgebra S of A exists such that $S/(rad(A)^2) = T$ is true. Hence, we deduce

$$S/(rad(A)^2) \oplus rad(A)/(rad(A)^2) = A/(rad(A)^2).$$

$rad(A)^2$ is a nilpotent ideal of A such that $S/(rad(A)^2)$ is separable. We conclude $rad(S) = rad(A)^2$. In addition, $S/rad(S)$ is isomorphic to T which is separable. A is unitary, and thus B is unitary, too. We use remark 4 to deduce that T is unital within B. S is the epimorphic pre-image of the natural epimorphism from A to $A/(rad(A)^2)$. Thus, $S = \{a \in A \mid$

$a + rad(A)^2 \in T\}$. T is unital, and thus $1_A + rad(A)^2 \in T$. We deduce $1_A \in S$. T is G-invariant. We deduce $(S/(rad(A)^2))g \leq S/(rad(A)^2)$ for all $g \in G$. Therefore $S.g \leq S + rad(A)^2 = S$ for all $g \in G$ is valid: S is G-invariant, too. We can apply induction on S because the nilpotency class of $rad(A)^2$ is smaller than the one of $rad(A)$. Thus, a G-invariant subalgebra C exists such that $S = C \oplus rad(A)^2$. In addition,

$$S/(rad(A)^2) \oplus rad(A)/(rad(A)^2) = A/(rad(A)^2)$$

is valid. We prove $A = C \oplus rad(A)$. C is isomorphic $S/(rad(A)^2)$ which is T. The algebra T is isomorphic to $B/rad(B)$ which is isomorphic to $A/rad(A)$.◇

C.2 The conjugacy part

This section is dedicated to the proof of the uniqueness part of Taft's theorem based on Taft's presentation within [41], [42], [43], [44], [45] and [46].

C.2.1 The case of a zero nilradical - group cohomology

We use the strategy of the proof presented by Taft within[41]. Let S, T be two G-invariant radical complements in A and $rad(A)^2 = 0$. **We do not need the assumption** $char(K) \neq 2$ **in this case!** Let $z \in rad(A)$ such that $S\kappa_{1-z} = T$ is valid based on the Wedderburn-Malcev theorem. Let $g \in G$. At first we prove that $z - zg \in rad(A) \cap Z(A)$ is valid. If g is an automorphism, then we calculate based on the rules within subsection C.1.1:

$$\begin{aligned} S\kappa_{1-z} &= \\ T &= \\ Tg &= \\ (S\kappa_{1-z})g &= \\ S\kappa_{1-zg}. & \end{aligned}$$

Thus, $(1-z)(1-zg)^{-1}$ is normalizing S. By using remark 12 we deduce that this element is centralizing S. $rad(A)^2 = 0$, and based on the rules mentioned before the element is exactly

$$(1-z)(1+zg) = 1 - z + zg.$$

Thus $z - zg$ is centralizing S. Again, by using $rad(A)^2 = 0$ the element $z - zg$ is centralizing $rad(A)$, and we conclude $z - zg \in rad(A) \cap Z(A)$. Now let g be an anti-automorphism. In this case the argumentation is similar, and we prove that

$$(1-z)(1-zg)^{-1} = (1-z)(1+zg) = 1 - z + zg$$

is normalizing S. Again, we deduce $z - zg \in rad(A) \cap Z(A)$. We focus on the map

$$f : G \longrightarrow rad(A) \cap Z(A), g \mapsto z - zg.$$

Now we prove that f is a 1-cocycle of G into $rad(A) \cap Z(A)$. Let $g, h \in G$. We consider four cases:

(i) g, h are automorphism: $f(gh) = z - zgh = (z - zg)h + z - zh = f(g)h + f(h)$.

(ii) g is an auto- and h is an anti-automorphism: $f(gh) = z - z.(gh) = z + zgh = -zh + zgh + z + zh = (z - zg).h + z - z.h = f(g).h + f(h)$

(iii) g is an anti- and h is an automorphism: $f(gh) = z + zgh = (z + zg)h + z - zh = f(g).h + f(h)$

(iv) g, h are anti-automorphism: $f(gh) = z - zgh = -zh - zgh + z + zh = (z + zg).h + z + zh = f(g).h + f(h)$.

As presented within subsection C.1.1 the first cohomology group of G is zero. Thus, f is a 1-coboundary of G into $rad(A) \cap Z(A)$. By definition an element $x \in rad(A) \cap Z(A)$ exists such that for all $g \in G$ the identity $z - zg = x - xg$ is valid. Hence, $z - x$ is G-skew, and by using definition and remark 4 we conclude that $1 - z + x$ is G-orthogonal. We calculate by using $x \in Z(A)$, $rad(A)^2 = 0$ and the rules mentioned within subsection C.1.1:

$$T = S\kappa_{1-z} = (S\kappa_{1-z})^{1+x} = S\kappa_{(1-z)(1+x)} = S\kappa_{1-z+x}.$$

This calculation finishes this subsection.◊

C.2.2 The case of a zero nilradical - cohomology of algebras

The author mentions here that the same strategy as used within subsection C.1.2 can be applied again to avoid cohomology of groups but to use cohomology of associative algebras.◊

C.2.3 The induction argument - cohomology of groups

The cohomology approach of groups is related to the proof within [41] and [42]. Let S, T be two G-invariant radical complements of A. We consider the factor algebra $A/rad(A)^2$. The factor algebra is G-invariant, and $(S + rad(A)^2)/rad(A)^2$ resp. $(T + rad(A)^2)/rad(A)^2$ are two G-invariant radical complements of $A/rad(A)^2$ which are isomorphic to S resp. T. In addition, the nilradical of $A/rad(A)^2$ is $rad(A)/rad(A)^2$ which is a zero algebra. Thus, we can use subsection C.2.1 to derive the existence of an element $\hat{x} = x + rad(A)^2 \in rad(A)/rad(A)^2$ such that $\hat{x}$ is G-skew and

Proof of Taft's theorem for associative unitary algebras

$$(S + rad(A)^2)/rad(A)^2 \kappa_{1-\hat{x}} = (T + rad(A)^2)/rad(A)^2.$$

Let us define the function

$$f_1 : G \longrightarrow rad(A)^2, g \mapsto \tfrac{1}{2}(x.g - x).$$

The image of f_1 is contained in $rad(A)^2$ because $\hat{x}$ is G-skew. As proven case by case within subsection C.2.1 we can deduce that f_1 is a 1-cocycle, and thus it is a 1-coboundary. Therefor an element $z \in rad(A)^2$ exists such that

$$z - z.g = \tfrac{1}{2}(x - g - x)$$

is valid for all $g \in G$. We define

$$y_1 := -z - \tfrac{1}{2}x.$$

It is straightforward to check that y_1 is G-skew. In addition, we define

$$\hat{y}_1 := y_1 + rad(A)^2 = -\tfrac{1}{2}\hat{x}$$

and

$$u_1 := -2y_1(1 - y_1)^{-1}.$$

Within $rad(A)/rad(A)^2$ we calculate

$$\hat{u}_1 := u_1 + rad(A)^2 = -2(-0.5\hat{x})(1 + 0.5\hat{x})^{-1} = \hat{x}(1 - 0.5\hat{x}) = \hat{x}.$$

In addition,

$$1 - u_1 = 1 + 2y_1(1 - y_1)^{-1} = (1 + y_1)(1 - y_1)^{-1}$$

is valid. y_1 is G-skew, and thus $(1 + y_1)(1 - y_1)^{-1}$ is G-orthogonal. In particular,

$$(S + rad(A)^2)/rad(A)^2 \kappa_{1-\hat{x}} = \kappa_{1-\hat{u}_1} = (T + rad(A)^2)/rad(A)^2$$

and therefor

$$S\kappa_{1-u_1} \oplus rad(A)^2 = T \oplus rad(A)^2 =: C$$

is valid. Within the algebra C we can use induction on the nilpotency class. Hence, a G-orthogonal element $1 - r \in 1 - rad(A)^2$ exists such that $S\kappa_{1-u_1}\kappa_{1-r} = T$ is valid. We conclude, that S and T are conjugated by the G-orthogonal element $(1 - u_1)(1 - r)$.⋄

C.2.4 The induction argument - direct computation

The proof stated here is based to the one presented by Taft in [45]. Let S, T be two G-invariant radical complements of A. We consider the factor algebra $A/rad(A)^2$. The factor algebra is G-invariant, and $(S + rad(A)^2)/rad(A)^2$ resp. $(T + rad(A)^2)/rad(A)^2$ are two G-invariant radical complements of $A/rad(A)^2$ which are isomorphic to S resp. T. In addition, the nilradical of $A/rad(A)^2$ is $rad(A)/rad(A)^2$ which is a zero algebra. Thus, we can use subsection C.2.1 to derive the existence of an element $\hat{x} = x + rad(A)^2 \in rad(A)/rad(A)^2$ such that $\hat{x}$ is G-skew and

$$(S + rad(A)^2)/rad(A)^2 \kappa_{1-\hat{x}} = (T + rad(A)^2)/rad(A)^2.$$

We define

$$y := \frac{1}{|G|} \sum_{t \in G} (1)^{sgn(t)}(x.t)$$

in which $sgn(t) := 1$ resp. $sgn(t) := -1$ if g is an auto- resp. anti-automorphism. It is straightforward to check that y is a G-skew element of $rad(A)$. $\hat{x}$ is G-skew, and thus $\hat{y} := y + rad(A)^2 = \hat{x}$. In particular,

$$(S + rad(A)^2)/rad(A)^2 \kappa_{1-\hat{x}} = \kappa_{1-\hat{y}} = (T + rad(A)^2)/rad(A)^2$$

and

$$S\kappa_{1-y} \oplus rad(A)^2 = T \oplus rad(A)^2 =: C$$

are valid. Within the algebra C we can use induction on the nilpotency class. Hence, a G-orthogonal element $1 - z \in 1 - rad(A)^2$ exists such that

$$S\kappa_{1-y}\kappa_{1-z} = T$$

is valid. We conclude, that S and T are conjugated by the G-orthogonal element $(1-y)(1-z)$.⋄

We remark that we do not need the assumption of characteristic not equal to two. Therefor the uniqueness part can be generalized accordingly.

C.3 Exercises

Excercise 394 *Prove the induction argument of subsection C.1.4 again by starting the proof dual with $rad(A)^{cl(rad(A))-1}$ instead of using $rad(A)^2$.*

Excercise 395 *Generalize the existence part proven in this chapter in the following way: the conditions that $A/rad(A)$ is separable is replaced by the condition that A possesses a radical complement. For proving this generalized version analyze the proof presented in this chapter in details.*

Excercise 396 *Clarify within exercise 395 that for the trivial group no new proof of the Wedderburn-Malcev theorem is presented.*

Excercise 397 *Prove the rules mentioned in subsection C.1.1 in details.*

Excercise 398 *Prove within the context of the existence part of Taft's theorem that every G-invariant separable subalgebra can be conjugated orthogonally into a G-invariant radical complement.*

Excercise 399 *Prove corollary 1 within [46]: Within the context of the uniqueness part of Taft's theorem in characteristic zero the G-orthogonal conjugacy can be written in the form $exp(ad(x))$ where x is a G-symmetric element.*

Excercise 400 *Analyze the examples mentioned by Taft within [46] in chapter 6 (pages 27 to 29) and calculate the results in details.*

Excercise 401 *Execute the proof presented in subsection C.1.2 in details.*

List of Figures

the algebras of upper and lower triangular matrices 25
multiplication table of a 4-dimensional algebra 31
an algebra without possessing a radical complement 32
multiplication table of a 3-dimensional algebra 32
an algebra possessing a separable factor algebra by its nilradical . 34

factor algebra by the nilradical of an adjunction of an unit 63
structure constants of a special 2-dimensional non-unitary associative algebra 71
radical complements of a special 2-dimensional non-unitary associative algebra 72
compatibility of the Wedderburn-Malcev theorem with ideals and factor algebras 79

solvable classes related to triangular matrices 115
a summarizing example for solvable associative algebras 121

multiplication matrix of $A(a,b,K)$ 137
multiplication matrix of $A(b,a,K)$ 138
multiplication matrix of $A(a,-1,K)$ 138
multiplication matrix of $A(a,c^2b,K)$ 138
$A(a,b,K)$ in characteristic not 2 141
infinite many intermediate fields 146
multiplication matrix of $A(c,d,K)$ 147
the algebras $A(a,b,K)$ in characteristic not 2 148

construction of the generalized Jordan decomposition 164
special subalgebras of commutative algebras 174
special subalgebras of $A(0,0,K)$ 175
special subalgebras of $A(a,0,K)$ 175
special subalgebras of $A(a,b,K)$ 176

Bibliography

[1] Eiichi Abe, Hopf Algebras, Cambridge University Press, 1977

[2] A. A. Baranov, A. Mudrov, H. M. Shlaka, Wedderburn-Malcev decomposition of one-sided ideals of finite-dimensional algebras, https://arxiv.org/abs/1609.05812, 2016

[3] Thorsten Bauer, Über die Struktur der Solomon-Algebren, Bayreuther Mathematische Schriften, Heft 63, 2001, 1-102

[4] Murray R. Bremner, How to compute the Wedderburn decomposition of a finite-dimensional associative algebra, https://pdfs.semanticscholar.org/0243/ea72b861f47bc1dbe17a570c13a612fcd1b0.pdf

[5] Sonia P. Coelho, C. Polcino Milies, Derivations of upper triangular matrix rings, Linear Algebra and its Applications, Volume 187, 1 July 1993, 263-267

[6] Charles W. Curtis, Irving Reiner, Representation theory of finite groups and associative algebras, Interscience Publishers, New York, London, 1962

[7] L. E. Dickson: Algebras and Their Arithmetics. University of Chicago Press, 1923

[8] Yurij A. Drozd, Vladimir V. Kirichenko, Finite dimensional algebras, Springer-Verlag, Berlin-Heidelberg, 1994

[9] Rolf Farnsteiner, The theorem of Wedderburn-Malcev: $H^2(A, N)$ and extensions, https://www.math.uni-bielefeld.de/ sek/select/RF6.pdf, 2005

[10] Rolf Farnsteiner, The theorem of Wedderburn-Malcev: conjugacy of maximal separable subalgebras, https://www.math.uni-bielefeld.de/ sek/select/RF7.pdf, 2005

[11] Francis J. Flanigan, A little Wedderburn principal theorem, Rocky Mountain journal of mathematics, volume 23, no. 1, pages 105-110, 1993

[12] George Glauberman, Fixed points in groups with operator groups, Math. Z. 84, 120-125, 1964

[13] W.A. de Graaf, G. Ivanyos, A. Küronya, L. Ronyai, Computing Levi Decomposition in Lie algebras, AAECC 8, 1997, 291-303

[14] G. Hochschild, Semi-simple algebras and generalized derivations. Amer. J. Math. 64 (1941), 677-694

[15] G. Hochschild, On the cohomology groups of an associative algebra. Ann. of Math. 46 (1945), 58-67

[16] G. Hochschild, On the cohomology theory for associative algebras. Ann. of Math. 47 (1946), 568-579

[17] L.K. Hua, Some properties of s-fields, Proc. Nat. Acad. Sci. U.S.A. 35, 1949, 533-537

[18] Bertram Huppert, Endliche Gruppen I, Springer-Verlag, Berlin, 1967

[19] Ted Hurley, Group rings and rings of matrices, International journal of pure and applied mathematics, volume 31, no. 3, 319-335, 2006

[20] N. Jacobon, Basic algebra II, second edition, Dover Publications, Inc. Mineola, New York, 2009

[21] Geoffrey Janssens, A glimpse into the asymptotics of polynimial identities, http://www.nieuwarchief.nl/serie5/pdf/naw5-2017-18-1-060.pdf

[22] Gregory Karpilovsky, The jacobson radical of group algebras, Elsevier, Amsterdam, 1987

[23] Gregory Karpilovsky, Unit groups of classical rings, Clarendon Press, Oxford, 1988

[24] M.K. Kopp, Fréchet algebras of finite type, Archiv der Mathematik, September 2004, volume 83, issue 3, 217-228

[25] Max-Albert Knus et al, The book of involutions, AMS Colloquium Publications, Volume 44, 1998

[26] Irwin Kra, Santiago S. Simanca, On circulant matrices, Notices of the AMS, volume 59, no. 3, 03/2012

[27] Herbert Kupisch, Einreihige Algebren über einem perfekten Körper, Journal of Algebra 33, 68-74, 1975

[28] Hartmut Laue, Algebra II, Lecture Notes am Mathematischen Seminar der Christian-Albrechts-Universität zu Kiel, Sommersemester 2012

[29] Hartmut Laue, Assoziative Algebren, Vorlesung am Mathematischen Seminar der Christian-Albrechts-Universität zu Kiel, Wintersemester 1997/1998

[30] Hartmut Laue, Introduction to algebra, Mathematischen Seminar der Christian-Albrechts-Universität zu Kiel, 2016

[31] A. Malcev, On the represantation of an algebra as a direct sum of the radical and a semi-simple subalgebra, C. R. (Doklady) Acad. Sci. URSS (N.S.) 36, 1942, 42-45

[32] I. B. S. Passi, D. S. Passman, S. K. Sehgal, Lie solvable group rings, Can. J. Math., Vol. XXV, No. 4, 1973, 748-757

[33] D. S. Passman, Observations on group rings, Communications in Algebra, 5(11), 1977, 1119-1162

[34] S. Perlis, G.L. Walker, Abelian group algebras of finite order, Trans. Amer. Math. Soc. 68, 1950, pp. 420-426

[35] Richard Pierce, Associative algebras, Springer-Verlag, New York, 1982

[36] K. W. Roggenkamp, Cohomology of Lie-algebras, groups and algebras, Seminar Series in Mathematics, Algebra 1, Ovidius university, Mai 1994.

[37] Günter Scheja, Uwe Storch, Lehrbuch der Algebra, Teil 2, B.G. Teubner, Stuttgart, 1988

[38] Ian Stewart, Martin Golubitsky, Coordinate Changes for Network Dynamics, preprint, June 15, 2015

[39] Dimitrij A. Suprunenko, Matrix groups, AMS, Providence, Rhode Island, 1976, Translations of Mathematical Monographs, Volume 45

[40] Fernando Szechtmann, Jordan-Chevalley decomposition and the Wedderburn-Malcev theorem, Master of science, University of Alberta, 1994

[41] Earl J. Taft, Cohomology of algebraic groups and invariant splitting of algebras, Bull. Amer. Math. Soc. Volume 73, Number 1, 106-108, 1967

[42] Earl J. Taft, Cohomology of groups of algebra automorphism, Journal of algebra 10, 400-410, 1968

[43] Earl J. Taft, Invariant Wedderburn factors, Illinois J. Math. 1, 565-573, 1957

[44] Earl J. Taft, Uniqueness of invariant Wedderburn factors, Illinois J. Math. 6, 353-356, 1962

[45] Earl J. Taft, Cayley symmetries in associative algebras, Canad. J. Math. 15, 285-290, 1963

[46] Earl J. Taft, Orthogonal conjugacies in associative and Lie algebras, Trans. Amer. Math. Soc. 113, 18-29, 1964

[47] Earl J. Taft, Orthogonal Conjugacies in Finite Groups, Math. Annalen 170, 37-40, 1967

[48] Peter J. Webb, An introduction to the cohomology of groups, www-users.math.umn.edu/ webb/.../8246CohomologyNotes.pdf

[49] J.H.M. Wedderburn, On hypercomplex numbers. Proc. London Math. Soc. 6 (1908), 77-118

[50] David J. Winter, Abstact Lie algebras, Cambridge, Massachusetts, London, England, 1972

[51] Sven Wirsing, Endvertauschbare Anordnungen und die Struktur der Einheitengruppen modularer Gruppenalgebren, disserta-Verlag, Hamburg, 2015

[52] Sven Wirsing, Über die Struktur der Solomon-Tits-Algebren der symmetrischen Gruppen, disserta-Verlag, Hamburg 2015

[53] Sven Wirsing, Maximal nilpotente Teilstrukturen I, disserta-Verlag, Hamburg, 2015

[54] Sven Wirsing, Maximal nilpotente Teilstrukturen II, disserta-Verlag, Hamburg, 2016

[55] Sven Wirsing, Maximal nilpotent subalgebras I, anchor academic publishing, Hamburg, 2016

[56] Sven Wirsing, Maximal nilpotent subalgebras II, anchor academic publishing, Hamburg, 2017

[57] Sven Wirsing, Separabilität in kommutativen und auflösbaren Algebren, disserta-Verlag, Hamburg 2015

[58] Joo Heon Yoo, The Jordan-Chevalley decomposition, math.uchicago.edu/ may/REU2014/REUPapers/Yoo.pdf, 2014

[59] Fuzhen Zhang, Matrix-Theorie, Springer-Verlag, New York, 1999

[60] http://math.stackexchange.com/questions/1006540

[61] http://wikipedia.org/wiki/Companion matrix

[62] http://mathoverflow.net/questions/208713/irreducible-characters-of-finite-Abelian-groups

[63] https://mathoverflow.net/questions/71869/an-r-algebra-a-is-r-separable-if-and-only-if-all-derivations-are-inner

[64] https://mathoverflow.net/questions/309753/operation-of-a-p-group-on-a-set-of-p-power-order-and-fix-points

Index

About a theorem of Thorsten Bauer, 215
adjunction of an unit
   definition, 58
   elementary properties, 58
   existence of radical complements, 64
   factor algebra by the nilradical, 63
   factor and subalgebras, 59
   nilradical, 63
   separability, 63
   star composition, 68
   unitary algebras, 60
   Wedderburn-Malcev, 64
algebra
   group of automorphism, 18
   set of anti-automorphism, 18
associative algebra
   solvability and circle group, 217
   solvability of its group of units, 216

characteristical polynomial, 115
commutative algebras, 151
   basis for $D(A)$, 178
   calculating minimal polynomials, 169
   center, 151
   construction generalized Jordan decomposition, 163
   context of the Wedderburn-Malcev theorem, 157
   decomposition of the group algebra, 179
   diagonalizable, 154
   fully separable, 154
   generalized Jordan decomposition, 160
   group algebra, 176
   Hasse diagram, 174
   nilpotent, 154
   non-unitary, 180
   primitive idempotents in group algebras, 178
   properties generalized Jordan decomposition, 167
   semisimple, 154
   separable, 154
   splitting, 154
   the subalgebra $VSep(A)$, 154
   the subalgebras $ZF(A)$ and $D(A)$, 172

derivation
   1-coboundary, 17
   1-cocycle, 17
   definition, 17
   examples, 18
   first Hochschild cohomology group, 17
   inner derivation, 17
   inner derivations of separable algebras, 22
   separable algebra, 20
   Solomon algebras, 23
   upper triangular matrices, 26

Euler's totient function, 179

factor sets
   1-coboundary, 27
   2-cocycle, 27

definition, 27
second Hochschild cohomology group, 27
separable algebra, 28
splitting, 27

generalized quaternion algebras
  commutativity, 141
    structure of $A(a,0)$, 143
    structure of $A(a,b)$, 144
  definition, 137
  isomorphism, 137
  structure of $A(0,0)$, 139
  structure of $A(a,0)$, 140
  structure of $A(a,b)$, 140
Gerhard Paul Hochschild, 17
group
  1-coboundary, 230
  1-cocycle, 230
  first cohomology group, 230
  orbit length, 73
  set of right cosets, 73

Helmut Hasse, 174

Leonard Eugene Dickson, 104
Leopold Kronecker, 161

minimal polynomial, 115

Proof of Taft's theorem for associative unitary algebras, 229
Proof of the Wedderburn-Malcev theorem for associative unitary algebras, 219

radical complements
  G-invariant, 41
  G-orthogonal, 40
  G-skew, 40
  G-symmetric, 40
  algorithm, 80
  cardinality, 74
  compatability with ideals and factor algebras, 78

definition, 29
lifting, 80
non-existence, 31
normalizer and centralizer, 73
star-invariant, 76
tensor products, 65
top down calculation, 80
triangular matrices, 34
unital, 30
Wedderburn-Malcev, 29

semisimple algebra
  unitarity, 61
separable algebra
  characterization, 14
  cohomology, 17
  compatability with semisimple algebras, 65
  compatability with structures, 62
  definition, 14
  example, 14
  group algebra, 15
  matrix algebra, 16
  properties, 14
set of all nilpotent elements, 103
shift, 66
Solomons algebra
  center, 154
solvable associative algebras
  a summarizing example, 119
  algebraical closed, 122
  bilinear form, 103
  chain of derived subalgebras, 97
  chains of substructures, 94
  characterization by bilinear forms, 106
  class of solvability, 97
  classes of solvability for triangular matrices, 112
  compatabilities, 95
  compatability with sub- and factor algebras, 117
  definition, 93
  fully separable elements, 188

fully separable elements and radical complements, 190
fully separable elements as radical complement, 190
group algebras, 110
importance for associative algebras, 123
intersection with radical complements, 117
Lie characterization, 101
linking radical complement and Jordan decomposition, 164
nilradical, 115
semisimple subalgebras, 115
solvable group algebras, 110
solvable radical, 96
solvable residuum, 96
splitting field, 116
standard trace form, 105
tensor product, 102
triangular matrices, 111

star composition
G-orthogonal, 76
G-skew, 76
G-symmetric, 76
adjunction of an unit, 68
conjugation, 68
definition and properties, 67
ideal, 78
radical complements, 68
structure of separable subalgebras, 70
Wedderburn-Malcev, 68

theorem of Stuth, 215
triangular matrices
center, 154

Wilhelm Karl Joseph Killing, 107